ESSENTIAL MOLECULAR BIOLOGY REVIEW

ESSENTIAL MOLECULAR BIOLOGY REVIEW

PHILIP W. HALL III, MD
Associate Dean for Curriculum
Case Western Reserve University School of Medicine
Cleveland, Ohio

b
Blackwell Science

BLACKWELL SCIENCE

EDITORIAL OFFICES:
238 Main Street, Cambridge, Massachusetts 02142, USA
Osney Mead, Oxford OX2 0E1, England
25 John Street, London WC1N 2BL, England
23 Ainslie Place, Edinburgh EH3 6AJ, Scotland
54 University Street, Carlton, Victoria 3053, Australia
Arnette Blackwell SA, 224, Boulevard Saint Germaine, 75007 Paris, France
Blackwell Wissenschafts-Verlag GmbH Kurfürstendamm 57, 10707 Berlin, Germany
Zehetnergasse 6, A-1140 Vienna, Austria

DISTRIBUTORS:

USA
Blackwell Science, Inc.
238 Main Street
Cambridge, Massachusetts 02142
(Telephone orders: 800-215-1000 or 617-876-7000)

Canada
Copp Clark, Ltd.
2775 Matheson Blvd. East
Mississauga, Ontario
Canada, L4W 4P7
(Telephone orders: 800-263-4374 or 905-238-6074)

Australia
Blackwell Science Pty., Ltd.
54 University Street
Carlton, Victoria 3053
(Telephone orders: 03-9347-0300
Fax 03-9349-3016)

Outside North America and Australia
Blackwell Science, Ltd.
c/o Marston Book Services, Ltd.
P.O. Box 269
Abingdon
Oxon OX14 4YN
England
(Telephone orders: 44-01235-465500
Fax: 44-01235-465555)

Acquisitions: Joy Denomme
Development: Kathleen Broderick
Production: Irene Herlihy
Manufacturing: Karen Feeney
Typeset by Dartmouth Publishing, Inc.
Printed and bound by DS Graphics

Printed in the United States of America
96 97 98 99 5 4 3 2 1

CONTENTS

---

1. SUBJECT AREA: The Evolution of the Cell

QUESTION: In experiments set up to imitate the conditions of early life, which type of cellular organic molecules were not formed?

A: amino acids
B: fatty acids
C: nucleotides
D: sugars
E: each of the above are formed.

Learning Response E: Correct. In addition to forming a broad range of simpler organic molecules, each of the above are formed by experiments using a mixture of gases as source for C, H, N, and O.

---

2. SUBJECT AREA: The Evolution of the Cell

QUESTION: When polynucleotides direct their own synthesis, they form a molecule with:

A: complementary sequence
B: identical sequence
C: polynucleotides cannot direct their own synthesis
D: random sequence
E: none of the above are formed.

Learning Response A: Correct. Through the specific base pairing formed by nucleotides, a chain of complementary sequence is formed. (However in subsequent rounds, this first product in turn makes a complement identical to the original.)

---

3. SUBJECT AREA: The Evolution of the Cell

QUESTION: In the primordial environment, the selective preservation of a change in the sequence of an RNA molecule might have occurred for any of the following reasons EXCEPT:

A: the ability of the molecule to assemble DNA.

B: the ability of the molecule to catalyze a reaction.
C: the ability of the molecule to fold in solution.
D: the efficiency of the molecule to replicate.
E: all of the above would have been selected for in the primordial environment.

Learning Response A: Correct. While it has since established a major role, DNA is not thought to have played a role in selection of RNA in the primordial environment.

---

4. SUBJECT AREA: The Evolution of the Cell
QUESTION: Evolution involves the following, EXCEPT:

A: descent from a common ancestor.
B: selection in favor of specific traits.
C: spontaneous appearance of novel organisms.
D: the occurrence of random variation in genetic makeup.
E: all of the above.

Learning Response C: Correct. Evolution is the process by which multiple species have derived from a common ancestor through gradual divergence. This involves a selective advantage for certain inherited traits over others.

---

5. SUBJECT AREA: The Evolution of the Cell
QUESTION: DNA is more stable for storage of genetic information than RNA because:

A: being double stranded, it is more rigid.
B: containing thymine, it is less susceptible to chemical damage than RNA with uracil.
C: having fewer hydroxyl groups, it is less hydrophobic.
D: having two strands, each serves a template to repair damage to the other.
E: not being catalytic, it is not involved in chemical reactions.

Learning Response D: Correct. Because of the differing chemical structure of its sugar, DNA is less susceptible to hydrolysis than RNA, as well as having the ability to be repaired when such damage occurs.

---

6. SUBJECT AREA: The Evolution of the Cell

QUESTION: Each of the following are examples of procaryotes EXCEPT:

A: archaebacteria.
B: cyanobacteria.
C: mycoplasma.
D: spirochetes.
E: trypanosoma.

Learning Response E: Correct. Trypanosomes are protists, single celled eucaryotes.

---

7. SUBJECT AREA: The Evolution of the Cell

QUESTION: Certain cells perform photosynthesis, which involves:

A: the degradation of oxygen to produce water.
B: the fixation of atmospheric nitrogen.
C: the generation of high energy electrons.
D: the interaction of sunlight with chloroform.
E: none of the above.

Learning Response C: Correct. Photosynthesis involves the excitation of chlorophyll by sunlight, which produces high energy electrons. These are passed to oxygen, which is produced through the breakdown of water, and results in the conversion of $C0_2$ to organic carbon.

---

8. SUBJECT AREA: The Evolution of the Cell

QUESTION: Which of the following eucaryotic organelles is NOT surrounded by a double membrane:

A: chloroplasts.
B: mitochondria.

C: the Golgi apparatus.
D: the nucleus.
E: all of the above have double membranes.

Learning Response C: Correct. The golgi are membrane bound, but by only a single membrane.

---

9. SUBJECT AREA: The Evolution of the Cell
QUESTION: All of the following include a lipid bilayer EXCEPT:

A: a centriole.
B: a vacuole.
C: peroxisomes.
D: the endoplasmic reticulum.
E: the plasma membrane.

Learning Response A: Correct. The centrioles, located in the cytoplasm, are not directly associated with a membrane.

---

10. SUBJECT AREA: The Evolution of the Cell
QUESTION: Procaryotes and eucaryotes are similar in that both:

A: are primarily anaerobic in their metabolism.
B: contain extensive cytoskeletons.
C: have similar chromosome structures.
D: typically contain organelles.
E: None of the above characteristics are shared.

Learning Response E: Correct. These two divisions of life, while sharing many of the basic biochemical reactions, have little else in common, when generalizing across groups.

---

11. SUBJECT AREA: The Evolution of the Cell
QUESTION: The elements of the eucaryotic cytoskeleton are directly responsible for each of the following processes EXCEPT:

A: DNA partitioning during cell division.
B: flagellar movement.
C: muscle contraction.
D: organelle tracking.
E: RNA translation.

Learning Response E: Correct. While actin filaments, microtubules, and other cytoskeletal elements are involved in many cellular processes, translation of RNA does not appear to be one of them, instead involving ribosomes, and sometimes the endoplasmic reticulum.

---

12. SUBJECT AREA: The Evolution of the Cell
QUESTION: All protozoa contain:

A: chloroplasts.
B: cilia.
C: flagella.
D: mitochondria.
E: multiple cells.

Learning Response D: Correct. While some protozoa possess flagella, cilia or chloroplasts, mitochondria are necessary for all eucaryotes, including protozoa.

---

13. SUBJECT AREA: The Evolution of the Cell
QUESTION: Which of the following cell types contains a single diploid nucleus in each cell?

A: nerve cell.
B: red blood cell.
C: skeletal muscle cell.
D: sperm cell.
E: all eucaryotic cells contain a diploid nucleus.

Learning Response A: Correct. Of those listed, only the nerve cell contains a single diploid nucleus. Red blood cells have no nucleus, skeletal muscle cells have many nuclei per cell, while a sperm cell is haploid.

14. SUBJECT AREA: The Evolution of the Cell
QUESTION: Each of the following are secreted by their respective cell EXCEPT:

A: extracellular matrix by fibroblasts.
B: extracellular matrix by osteoblasts.
C: hormones by gland cells.
D: lipid by adipose cells.
E: all of the above are secreted by the respective cells.

Learning Response D: Correct. Lipid is stored by adipose cells within intercellular, not extracellular droplets.

---

15. SUBJECT AREA: The Evolution of the Cell
QUESTION: Which of the following is a property of eucaryotic DNA?

A: all DNA encodes protein.
B: different regions of DNA are transcribed in different cell types.
C: DNA is kept from wrapping up into compact structures by histones.
D: the amount of DNA is proportional to the complexity of the organism.
E: none of the above are true of eucaryotic DNA.

Learning Response B: Correct. Most eucaryotic DNA does not encode proteins, with much of the unused DNA kept compact by histones. There appears to be no relationship between the amount of DNA and the complexity of an organism. However, the differences between cell types are in part due to their expression of different proteins.

---

16. SUBJECT AREA: The Evolution of the Cell
QUESTION: A plasma membrane:

A: does not form under physiologic conditions.
B: is a monolayer.

C: is comprised primarily of polypeptides.
D: is comprised of amphipathic molecules.
E: separates the hydrophobic inside of a cell from the hydrophilic outside.

Learning Response D: Correct. Plasma membrane is a bilayer comprised of amphipathic lipid molecules which separate two hydrophilic environments.

---

17. SUBJECT AREA: The Evolution of the Cell
QUESTION: The cells of a *Hydra*:

A: are arranged randomly.
B: are not bound together mechanically.
C: each develops independently.
D: each performs a specialized function.
E: form as a single layer.

Learning Response D: Correct. The *Hydra* is a simple example of a multicellular organism because its cells use intercellular communication to govern their development. They form into two separate layers with different specialized functions.

---

18. SUBJECT AREA: The Evolution of the Cell
QUESTION: Developmental devices in higher organisms:

A: are distinct from those observed in lower organisms.
B: are not affected by evolution.
C: change rapidly through evolution.
D: have been conserved among related species.
E: none of the above.

Learning Response D: Correct. While many basic mechanisms have been conserved among all organisms, evolution has resulted in the development of characteristics which are unique to specific organisms. A high degree of similarity can still be observed among closely related animals.

19. SUBJECT AREA: The Evolution of the Cell
QUESTION: Germ cells:

A: are a specialized class of somatic cells.
B: are also known as zygotes.
C: are necessary for asexual reproduction.
D: serve to limit the recombination of genetic material.
E: usually occur in two different forms.

Learning Response E: Correct. Germ cells which are distinct from somatic cells usually occur as sperm and egg, which fuse to form a zygote. This form of sexual reproduction allows for greater genetic recombination.

---

20. SUBJECT AREA: The Evolution of the Cell
QUESTION: The cells involved at a neuromuscular junction:

A: are all of the same type.
B: are highly specialized.
C: have not been well characterized.
D: play similar roles.
E: none of the above.

Learning Response B: Correct. The neuromuscular junction has been studied as a model for the interaction between different types of cells. Each cell type is highly specialized and plays a unique role in the interaction.

---

21. SUBJECT AREA: The Evolution of the Cell
QUESTION: All of the following are true of the immune system EXCEPT:

A: it is comprised of a large number of different cell types.
B: it uses a process of variation and selection.
C: it uses unique genetic mechanisms to generate antibodies.

D: its primary task is to destroy foreign microorganisms in the body.
E: all of the above are true.

Learning Response E: Correct. The immune system uses a wide variety of cell types to form a defense against foreign proteins and microorganisms. Part of this process relies on unique genetic recombination to produce variability, followed by selection for immunogenicity.

---

22. SUBJECT AREA: The Evolution of the Cell
QUESTION: All of the following are true of the nervous system, EXCEPT:

A: involves the use of messenger molecules.
B: is present in almost all multicellular animals.
C: utilizes many diverse cell types.
D: works entirely by direct cell-cell interaction.
E: all of the above are true.

Learning Response D: Correct. The nervous system is present in most multicellular animals. Using numerous specialized cell types and messenger molecules, it allows communication either with adjacent cells or with distant points in the body.

---

23. SUBJECT AREA: The Evolution of the Cell
QUESTION: Connections among nerve cells:

A: always require sensory input to form appropriately.
B: are entirely the result of genetic factors.
C: are formed through mechanisms distinct from those used to perform similar functions in other cells.
D: constrain the patterns of animal behavior.
E: form randomly with any adjacent cell.

Learning Response D: Correct. Nerve cells use conserved mechanisms to form the connections which determine animal behavior. Through a combination of genetic and sensory input, specific contacts are made.

---

24. SUBJECT AREA: The Evolution of the Cell

QUESTION: Which of the following statements concerning eucaryotic organelles is FALSE?

I. some are thought to have arisen by symbiosis.
II. they are identical in all eucaryotes.
III. they carry out specialized functions for cells.
IV. they represent specialized cells.

A: I and III.
B: I and IV.
C: II and III.
D: II and IV.
E: I, II and III.

Learning Response A: Correct. Organelles which are special membrane bound regions within eucaryotic cells carry out specific functions for these cells. Some are thought to have been acquired through symbiosis including mitochondria and chloroplasts in plants.

---

25. SUBJECT AREA: Small Molecules, Energy, and Biosynthesis

QUESTION: Which of the following amino acids might disrupt a protein structure due to atypical linkage of side chain to the peptide chain?

A: alanine.
B: arginine.
C: glutamine.
D: proline.
E: tyrosine.

Learning Response D: Correct. Proline which involves a ring including the peptide backbone, does not allow free rotation around the α carbon.

---

26. SUBJECT AREA: Small Molecules, Energy, and Biosynthesis

QUESTION: Which amino acid can form disulfide bonds to cross-link two amino acid chains?

A: cysteine.
B: histidine.
C: methionine.
D: serine.
E: valine.

Learning Response A: Correct. While both methionine and cysteine contain sulfurs in their side chains, only that of cysteine is free to form covalent bonds.

---

27. SUBJECT AREA: Small Molecules, Energy, and Biosynthesis
QUESTION: Which of the following molecules are the primary component of cell membranes?

A: fatty acids.
B: polysaccharides.
C: proteins.
D: pyrimidines.
E: triglycerides.

Learning Response A: Correct. While proteins and cholesterol are found in membranes, fatty acids, with long carbon chains and polar head groups are the primary membrane components.

---

28. SUBJECT AREA: Small Molecules, Energy, and Biosynthesis
QUESTION: All of the following types of molecules can contain nucleotide as part of their structure EXCEPT:

A: amino acids
B: chemical energy sources
C: coenzymes
D: genetic material
E: signaling molecules

Learning Response A: Correct. The nucleotide adenosine is used as an energy source (ATP), as a signaling molecule (cAMP), and forms a part of coenzyme A. Nucleotides also constitute DNA and RNA but not amino acids.

29. SUBJECT AREA: Small Molecules, Energy, and Biosynthesis
QUESTION: Each of the following represents a major class of small molecules used by the cell EXCEPT:

A: amino acids.
B: fatty acids.
C: nucleotides.
D: steroids.
E: sugars.

Learning Response D: Correct. While all of these molecules can be found in cells, steroids are not utilized by cells to the large extent that the other molecules are.

---

30. SUBJECT AREA: Small Molecules, Energy, and Biosynthesis
QUESTION: Which of the following statements regarding sugars is FALSE:

A: they all have the structure $C_6H_{12}O_6$.
B: they are oxidized to produce energy.
C: they are used to form extracellular structures.
D: they can be found in large oligomers.
E: they can be linked to proteins or lipids.

Learning Response A: Correct. While $C_6H_{12}O_6$ is the structure of the sugar, glucose, not all sugars have this structure.

---

31. SUBJECT AREA: Small Molecules, Energy, and Biosynthesis
QUESTION: All the following can be associated with phospholipids EXCEPT:

A: cysteine.
B: choline.
C: ethanolamine.
D: serine.
E: all the above are associated with phospholipids.

Learning Response A: Correct. Unlike serine, cysteine is an amino acid which is not linked to phospholipids.

---

32. SUBJECT AREA: Small Molecules, Energy, and Biosynthesis
QUESTION: Which of the following are pyrimidines?

I. adenine.
II. cytosine.
III. guanine.
IV. thymine.
V. uracil.

A: I and III.
B: II and III.
C: II and IV.
D: I, III and V.
E: II, IV and V.

Learning Response E: Correct. The pyrimidines include cytosine, thymine, and uracil while the purines are adenine and guanine.

---

33. SUBJECT AREA: Small Molecules, Energy, and Biosynthesis
QUESTION: All the following are properties of amino acids EXCEPT:

A: some change charge with a change in pH.
B: they are grouped by the nature of their side chains.
C: they are joined by peptide bonds.
D: they form circular polymers.
E: they serve as subunits in the synthesis of proteins.

Learning Response D: Correct. Proteins consist of linear polymers of amino acids joined by peptide bonds. Properties of the protein depend on the properties of the amino acid side chains and include effects of pH induced changes in charge.

---

34. SUBJECT AREA: Small Molecules, Energy, and Biosynthesis
QUESTION: Hydrogen bonding:

A: involves polar interaction.
B: is as strong as covalent bonding.
C: occurs in molecular hydrogen.
D: prevents clustering of hydrophobic molecules.
E: prevents interaction with charged molecules.

Learning Response A: Correct. Hydrogen bonding requires a polar molecule. Such molecules have partially charged poles which can then interact with either charged molecules or other polar molecules. This tends to bring together hydrophobic molecules by exclusion.

---

35. SUBJECT AREA: Small Molecules, Energy, and Biosynthesis
QUESTION: The second law of thermodynamics tells us that:

A: disorder always increases.
B: disorder can only decrease in a system which is not isolated.
C: disorder only increases in an isolated system.
D: in a closed system disorder remains constant.
E: in a system which is not closed, disorder will always decrease.

Learning Response B: Correct. The only conditions under which disorder can (but need not) decrease is in a system which is not isolated. In a closed system, disorder always increases.

---

36. SUBJECT AREA: Small Molecules, Energy, and Biosynthesis
QUESTION: To compensate for the increase in order caused by certain biological functions, cells:

A: create energy.
B: decrease disorder.
C: release heat.
D: violate the second law of thermodynamics.
E: none of the above.

Learning Response C: Correct. Through a tight coupling of heat release to a disorder-decreasing reaction, cells productively utilize available energy.

---

37. SUBJECT AREA: Small Molecules, Energy, and Biosynthesis
QUESTION: The oxidation of biological molecules:

A: involves the addition of electrons.
B: is coupled to the breakdown of ATP.
C: is performed through successive minor reactions.
D: liberates molecular oxygen.
E: results in their destruction with no benefit to the cell.

Learning Response C: Correct. In the oxidative process, some of the energy released through the removal of electrons is used for the formation of ATP. These electrons are passed to oxygen through several steps producing water.

---

38. SUBJECT AREA: Small Molecules, Energy, and Biosynthesis
QUESTION: The process of glycolysis:

A: creates simple sugars from polysaccharides.
B: involves the breakdown of acetyl CoA.
C: is a three stage process.
D: produces NADH.
E: requires oxygen to produce ATP.

Learning Response D: Correct. Glycolysis refers to the process by which simple sugars such as glucose are broken down. By nine enzymatic reactions, it produces pyruvate, ATP and NADH even in the absence of oxygen.

---

39. SUBJECT AREA: Small Molecules, Energy, and Biosynthesis
QUESTION: The citric acid cycle requires all the following EXCEPT:

A: acetyl CoA.
B: ADP.
C: FAD.

D: GDP.
E: NAD.

Learning Response B: Correct. The citric acid cycle uses acetyl CoA as a substrate to produce NADH(from NAD), $FADH_2$(from FAD), and GTP(from GDP), all of which can serve as energy sources for the production of ATP. However, it does not use ADP directly.

---

40. SUBJECT AREA: Small Molecules, Energy, and Biosynthesis
QUESTION: The process of oxidative phosphorylation:

A: converts all the chemical energy of NADH into ATP.
B: involves a single enzyme.
C: is equivalent to burning oxygen to form water.
D: produces a proton gradient.
E: takes place in the cell nucleus.

Learning Response D: Correct. Oxidative phosphorylation takes electrons from NADH and $FADH_2$ and passes them through a series of molecules in the mitochondrial membrane to oxygen. This reaction which is equivalent to burning oxygen produces a proton gradient which enables some of the energy from the electron transfer to be converted into ATP.

---

41. SUBJECT AREA: Small Molecules, Energy, and Biosynthesis
QUESTION: All the following accurately describe free energy in a system EXCEPT:

A: it is determined by the equation G=H-TS.
B: it must always decrease in a closed system.
C: it must decrease for a reaction to occur spontaneously.
D: it takes into account entropy and enthalpy.
E: all the above are true.

Learning Response E: Correct. Free energy (G) represents the degree of order in a system as defined by H-TS (H= enthalpy, T=temperature, S=entropy). The second law of thermodynamics requires a decrease in free energy in a closed system and because of this it must decrease for a reaction to be spontaneous.

---

42. SUBJECT AREA: Small Molecules, Energy, and Biosynthesis
QUESTION: Which of the following statements about cellular metabolism is true?

A: each small molecule is modified in a single specific way.
B: it is a complex web of enzymatic reactions.
C: it is subject to slow irreversible feedback regulation.
D: it is very easy to perturb the flow of small molecules.
E: it occurs at the same rate independent of the condition of the cell.

Learning Response B: Correct. Cellular metabolism involves complex patterns of reactions subject to rapid and reversible feedback mechanisms and other forms of regulation. Since each molecule can be used by several different enzymes, it is difficult to produce significant perturbations in the flow of molecules.

---

43. SUBJECT AREA: Small Molecules, Energy, and Biosynthesis
QUESTION: Which of the following statements regarding compartmentalization is FALSE?

A: allows sequential reactions to occur more efficiently.
B: can involve interactions between cells.
C: occurs on many levels.
D: only occurs on a macroscopic level.
E: utilizes cellular compartments to perform complex reactions.

Learning Response D: Correct. Compartmentalization occurs at many levels from two enzymes associating with each other to communication between cells in an organism. By confining sequential reactions, this process enables these reactions to be performed more efficiently.

44. SUBJECT AREA: Small Molecules, Energy, and Biosynthesis
QUESTION: Metabolic enzymes can be regulated by which of the following mechanisms?

I. covalent modification of the enzyme.
II. stimulation by a product of a pathway.
III. stimulation by the substrate.
IV. the results of degradation of the pathway's product.

A: I and III.
B: II and IV.
C: I, II and IV.
D: I, III and IV.
E: I, II, III and IV.

Learning Response D: Correct. There are many influences which regulate metabolic enzymes including feedback inhibition, covalent modifications, and availability of substrate. However, this would result in an inhibition by the product of the pathway.

---

45. SUBJECT AREA: Small Molecules, Energy, and Biosynthesis
QUESTION: The role of ATP in biosynthetic reactions is:

A: to absorb the energy they produce.
B: to couple its hydrolysis with energetically unfavorable reactions.
C: to covalently modify the enzyme.
D: to remove phosphate from the reaction.
E: to serve as an electron donor.

Learning Response B: Correct. By coupling the energy loss which accompanies ATP hydrolysis with an energetically unfavorable reaction, the enzyme is able to perform the reaction with a net decrease in free energy. In certain reactions it can also serve as the substrate (synthesis of RNA) or as a coenzyme (donating a phosphate group).

---

46. SUBJECT AREA: Small Molecules, Energy, and Biosynthesis

QUESTION: All the following are examples of coenzymes EXCEPT:

A: ATP.
B: biotin.
C: cGMP.
D: coenzyme A.
E: S-adenosylmethionine.

Learning Response C: Correct. Coenzymes serve as donors of chemical groups including phosphate, carboxyl, acetyl, and methyl groups for ATP, biotin, coenzyme A, and S-adenosylmethionine respectively.

---

47. SUBJECT AREA: Small Molecules, Energy, and Biosynthesis
QUESTION: NADPH:

A: differs from NADH by having two additional phosphate groups.
B: donates a hydride ion to an oxidation reaction.
C: is often linked to the use of a proton from the solution.
D: performs identical reactions to NADH.
E: none of the above are true.

Learning Response C: Correct. Differing from NADH by the presence of a single phosphate group, NADPH takes part in analogous but different reactions. Along with a proton from solution, NADPH donates a hydride ion to an enzymatic reaction.

---

48. SUBJECT AREA: Small Molecules, Energy, and Biosynthesis
QUESTION: The NADH produced by glycolysis:

A: donates directly to the electron transport system.
B: is distinct from that produced by the citric acid cycle.
C: is transferred to the citric acid cycle.
D: passes electrons directly to oxygen.
E: represents wasted energy.

Learning Response A: Correct. The NADH produced in glycolysis joins with that produced by the citric acid cycle to enter the electron transport system and indirectly passes their electrons to oxygen.

---

49. SUBJECT AREA: Macromolecules: Structure, Shape, and Information
QUESTION: The interaction between macromolecules involves the formation of all the following EXCEPT:

A: covalent bonds.
B: hydrogen bonds.
C: ionic bonds.
D: van der Waals attraction.
E: all the above are involved.

Learning Response E: Correct. In addition to the weak noncovalent interactions that determine the closeness of fit between molecules, they can also be linked by covalent bonds such as disulfide bridges.

---

50. SUBJECT AREA: Macromolecules: Structure, Shape, and Information
QUESTION: Which of the following bonds is the weakest:

A: covalent bond.
B: disulfide bond.
C: hydrogen bond.
D: ionic bond.
E: van der Waals attraction.

Learning Response E: Correct. Van der Waals attraction represents the weakest form of noncovalent bond, but also has the longest bond length.

---

51. SUBJECT AREA: Macromolecules: Structure, Shape, and Information
QUESTION: The following characterize the interaction between two macromolecules EXCEPT:

A: it can be represented by an affinity constant.
B: it occurs through the formation of multiple weak bonds.

C: it provides for specificity in a reaction.
D: its dissociation is often diffusion-limited.
E: all the above are true.

Learning Response D: Correct. The interaction between macromolecules provides for the specificity of a reaction through the formation of multiple weak bonds. The affinity constant is a measure of the strength of this interaction, which is a function both of the association (which is often diffusion limited) and the disassociation.

---

52. SUBJECT AREA: Macromolecules: Structure, Shape, and Information
QUESTION: Errors in molecular recognition:

A: allow for evolution.
B: are not compensated for by the cell.
C: are not predictable.
D: can be prevented.
E: never occur.

Learning Response A: Correct. While unavoidable, the rate of errors can be predicted and cells have mechanisms to compensate for or correct most errors. However, a low error rate in DNA replication or repair allows for evolution to take place.

---

53. SUBJECT AREA: Macromolecules: Structure, Shape, and Information
QUESTION: DNA consists of:

A: a double strand of nucleotides paired by covalent bonding.
B: a helical structure with a phosphate core.
C: bases paired by hydrogen bonding.
D: parallel strands conferring genetic information.
E: the nucleotides adenine, cytosine, guanine and uracil.

Learning Response C: Correct. DNA is a double helix of anti-parallel strands. The nucleotides bases (A, C, G, and T) point inward and through hydrogen bonding pair with bases on the opposing strand.

---

54. SUBJECT AREA: Macromolecules: Structure, Shapc, and Information
QUESTION: The linear sequence of DNA:

A: can not account for the hereditary information of the DNA.
B: is identical in all organisms.
C: is made up of the five nucleotides.
D: is perfectly replicated with each cell division.
E: serves as template for synthesis of the opposing strand.

Learning Response E: Correct. The sequence of the four nucleotides confer the hereditary information. The small error rate in using one strand as template for the other accounts for the divergence among organisms.

---

55. SUBJECT AREA: Macromolecules: Structure, Shape, and Information
QUESTION: In eucaryotes, the sequence of nucleotides in DNA:

A: is collinear with the protein sequence.
B: is identical to the mRNA.
C: is linked to a ribose-phosphate backbone.
D: is replicated through DNA transcription.
E: none of the above are true.

Learning Response A: Correct. Through DNA transcription, a cell makes a ribose containing RNA molecule which is then modified before serving as template for proteins. It was the colinearity of nucleotide and amino acid sequence that led to the discovery of the genetic code.

---

56. SUBJECT AREA: Macromolecules: Structure, Shape, and Information
QUESTION: RNA splicing:

A: allows more versatility in making proteins.
B: happens after leaving the nucleus.

C: happens prior to transcription.
D: is limited to bacterial cells.
E: occurs after its use to make protein.

Learning Response A: Correct. Splicing occurs immediately after transcription and before the mRNA is transported out of the nucleus. It is used by eucaryotes to allow a greater variety of proteins to be made.

---

57. SUBJECT AREA: Macromolecules: Structure, Shape, and Information
QUESTION: All the following are true about the genetic code EXCEPT:

A: it codes for 64 amino acids.
B: it determines the protein to be synthesized.
C: it has been highly conserved during evolution.
D: it is degenerate.
E: it is read in triplets.

Learning Response A: Correct. The genetic code has been highly conserved during evolution. It uses nucleotide triplets to code for an amino acid sequence. While there are 64 possible triplet combinations, the degeneracy of the system results in 20 amino acids being used, with some amino acids coded for by several triplets.

---

58. SUBJECT AREA: Macromolecules: Structure, Shape, and Information
QUESTION: A single transfer RNA molecule:

A: can interact with many different amino acids.
B: contains a four nucleotide anticodon.
C: entails the use of nontraditional base pairing.
D: is used in the transcription process.
E: maintains its cloverleaf structure through covalent bonding.

Learning Response C: Correct. tRNAs have a unique three base codon on one end and bind a unique amino acid on the other. They form their characteristic structure through hydrogen bonding interactions including triple stranded structures. They are used for translation of mRNA to protein.

---

59. SUBJECT AREA: Macromolecules: Structure, Shape, and Information
QUESTION: Catalytic activity of RNA has been implicated in the following EXCEPT:

A: biological origins.
B: maturation of tRNA transcripts.
C: ribosome activity.
D: self splicing.
E: catalytic activity has been implicated in all of the above.

Learning Response E: Correct. The discovery of the catalytic nature of self splicing RNA has resulted in similar catalytic roles being predicted for the RNA found in association with ribosomes and tRNA transcripts.This in turn has led to theories of RNA-only enzymes in the evolutionary past.

---

60. SUBJECT AREA: Macromolecules: Structure, Shape, and Information
QUESTION: All the following are true about the amino acid sequence of a protein EXCEPT:

A: it can allow for specialized structures to form within a protein.
B: it is disrupted by denaturation of a protein.
C: it is responsible for the confirmation of the protein.
D: it is structurally different from RNA and DNA.
E: the amino acids are linked by polar peptide bonds.

Learning Response B: Correct. A protein is formed by peptide bonds between amino acids which make its structure different than DNA and RNA. The specific sequence is responsible for the structure of the protein including active sites of enzymes. While not disrupting the sequence of the protein, denaturation involves the disruption of the folding which occurred during protein synthesis.

---

61. SUBJECT AREA: Macromolecules: Structure, Shape, and Information
QUESTION: X-ray diffraction studies of proteins have revealed the following structures EXCEPT:

A: antiparallel β sheets.
B: β helices.
C: coiled-coils.
D: parallel β sheets.
E: all the above.

Learning Response B: Correct. These studies have demonstrated the existence of β sheets formed by parallel and antiparallel strands. They have also revealed α helices which sometimes interact with each other in a coiled-coil arrangement.

---

62. SUBJECT AREA: Macromolecules: Structure, Shape, and Information
QUESTION: Secondary structure refers to :

A: the interaction of identical subunits.
B: the interaction of protein and DNA.
C: the interaction of protein domains.
D: the interaction to form helices and sheets.
E: the sequence of amino acids.

Learning Response D: Correct. Primary structure refers to the order of amino acids. By interacting as sheets or helices these demonstrate secondary structure. Tertiary and quaternary structures refer to the interaction of these helices and sheets into domains, proteins and their assembly as subunits.

---

63. SUBJECT AREA: Macromolecules: Structure, Shape, and Information
QUESTION: The following properties are shared by polypeptide domains EXCEPT:

A: they are formed by the interaction of helices and sheets.
B: they are often similar among related proteins.
C: they can be recombined to form proteins with differing activities.
D: they can be slightly modified to produce novel proteins.
E: they each demonstrate their own catalytic activity.

Learning Response E: Correct. Polypeptide domains are formed by the interaction between units of secondary structure and can be altered and recombined to form novel proteins. They often serve a conserved function which may be catalytic or structural and this causes their conservation among proteins performing similar functions.

---

64. SUBJECT AREA: Macromolecules: Structure, Shape, and Information
QUESTION: The formation of large protein structures through assembly of small polypeptides involves which of the following factors:

A: because of its modular nature, the assembly can not be easily controlled.
B: by combining subunits it allows for a broader range of products.
C: it results in more errors in the synthesis of the structure.
D: it results in the formation of highly irregular assemblages.
E: it requires more genetic information.

Learning Response B: Correct. Through the combination of subunits such a system allows for greater variability with less genetic information. The assembly can be more easily controlled and errors excluded to form a more uniform product.

---

65. SUBJECT AREA: Macromolecules: Structure, Shape, and Information
QUESTION: The catabolite activator protein homodimer demonstrates the following structural features EXCEPT:

A: disulfide bonding.
B: primary structure.
C: quaternary structure.
D: secondary structure.
E: tertiary structure.

Learning Response A: Correct. Being a homodimer this protein demonstrates all structures from primary to quaternary. There are no disulfide linkages in this protein.

---

66. SUBJECT AREA: Macromolecules: Structure, Shape, and Information
QUESTION: Self assembly of protein subunits:

A: can not explain the generation of a product of fixed length.
B: forms regular geometric structures.
C: is disrupted by nucleic acid.
D: is only used for the construction of linear structures.
E: only occurs among identical subunits.

Learning Response B: Correct. Self assembly involves the formation of geometric structures (rods, spheres, sheets or others) from identical or different subunits. These sometimes also involve nucleic acids which in some cases define a fixed length.

---

67. SUBJECT AREA: Macromolecules: Structure, Shape, and Information
QUESTION: A ligand refers to:

A: the binding site for a protein.
B: the molecule which specifically interacts with a protein's binding site.
C: the polar bond which forms a specific interaction with the protein.
D: the portion of the protein molecule which provides for its structure.
E: the surface of a binding molecule.

Learning Response B: Correct. A ligand interacts through specific hydrogen bonds with the binding site of a protein.

---

68. SUBJECT AREA: Macromolecules: Structure, Shape, and Information
QUESTION: The first step in enzyme catalysis is:

A: alteration of electron distribution.
B: binding of substrate.
C: dissociation of the product.
D: formation of an unstable transition state.
E: regeneration of the active site.

Learning Response B: Correct. The first and most important step in an enzyme reaction is the binding of the substrate. This is often followed by a change in the electron distribution to create an unstable transition state allowing the reaction to occur. The enzyme then must release its product and be regenerated before beginning a subsequent reaction.

---

69. SUBJECT AREA: Macromolecules: Structure, Shape, and Information
QUESTION: Which of the following statements about enzyme reactions is true:

A: their combined rates can be increased through the formation of a multienzyme complex.
B: their reaction rate is only limited by substrate availability.
C: they eliminate the need for coupling to ATP hydrolysis.
D: they make chemical reactions energetically more favorable.
E: they proceed equally well in either direction.

Learning Response A: Correct. Enzymatic reactions often couple ATP hydrolysis to chemical reactions to allow them to proceed preferentially in one direction, however, they can not make them energetically more favorable. Formation of a multienzyme complex can increase their efficiency, but enzymes still have a limit to the speed at which they can perform reactions.

---

70. SUBJECT AREA: Macromolecules: Structure, Shape, and Information
QUESTION: All of the following are true of allosteric proteins EXCEPT:

A: are limited in the types of processes in which they can be involved.

B: can serve as ion pumps.
C: can use ion gradients to do work.
D: can utilize their changes in conformation to do work.
E: often use ATP as an energy source.

Learning Response A: Correct. Often using ATP as an energy source to induce a conformational change, allosteric proteins can perform a broad range of useful functions in a cell.

---

71. SUBJECT AREA: Macromolecules: Structure, Shape, and Information
QUESTION: The role that allosteric proteins play in metabolism is:

A: their binding of reaction products shifts them to an inactive conformation.
B: their binding to product causes them to become more active.
C: they are limited to unusual circumstances.
D: they change conformation when they perform their reactions.
E: they often bind the first feedback molecule more rapidly than subsequent molecules.

Learning Response A: Correct. Many feedback pathways in the metabolic process involve the binding of allosteric proteins to reaction products or other chemical signals. The binding of product shifts these enzymes into a less active or inactive conformation. Binding of subsequent molecules of inhibitor is often accelerated by the initial binding event.

---

72. SUBJECT AREA: Macromolecules: Structure, Shape, and Information
QUESTION: All of the following relate to the use of intercellular membranes to decrease reaction rates EXCEPT:

A: it benefits from compartmentalization of substrates and enzymes.
B: it is analogous to a mechanism used by DNA binding proteins.
C: it limits diffusion to two dimensions.

D: it works best for reactions which are not diffusion limited.
E: all the above are true.

Learning Response D: Correct. The use of intercellular membranes relies on compartmentalization and the limiting of diffusion to accelerate the rate of diffusion limited reactions, as with the diffusion limited binding of proteins to DNA.

---

73. SUBJECT AREA: How Cells Are Studied
QUESTION: All the following are true of light microscopy EXCEPT:

A: it can resolve all cellular organelles.
B: it is limited by the wave length of light.
C: it often involves the use of tissue fixation and sectioning.
D: it suffers from diffraction effects.
E: it was used to first view cells.

Learning Response A: Correct. While its capabilities are enhanced by sample preparation, light microscopy is still limited by the wave length of light. In part due to diffraction effects, it cannot resolve anything smaller than mitochondria.

---

74. SUBJECT AREA: How Cells Are Studied
QUESTION: All of the following are used in light microscopy to better resolve the features of the cell EXCEPT:

A: electronic image processing.
B: fluorescence microscopy.
C: magnetic lenses.
D: phase-contrast microscopy.
E: selective dyes.

Learning Response C: Correct. Light microscopy uses numerous techniques to enhance its capabilities, however magnetic lenses are used in electron microscopy.

---

75. SUBJECT AREA: How Cells Are Studied

QUESTION: Which of the following is involved in scanning electron microscopy?

I. a beam of electrons.
II. magnetic lenses.
III. extremely thin tissue sections.
IV. direct viewing of transmitted electrons.

A: I and II.
B: I and IV.
C: I, II and III.
D: I, II and IV.
E: I, II, III and IV.

Learning Response A: Correct. Scanning microscopy uses a beam of electrons focused by magnetic lenses to view three dimensional samples. This works through a detector which converts the pattern of electrons which it receives into a visible image.

---

76. SUBJECT AREA. How Cells Are Studied

QUESTION: The freeze-fracture technique involves which of the following?

A: cryoprotectants to prevent ice crystal formation.
B: fixation of cells.
C: freezing to the temperature of liquid helium.
D: sublimation of ice after fracturing cells.
E: phase-contrast microscopy.

Learning Response A: Correct. Like the related freeze-etching technique, freeze-fracture uses scanning microscopy to visualize replicas of fractured cell surfaces. Because replicas are used, fixation is not necessary. It is the freeze etching that requires the use of liquid helium and sublimation, and does not allow cryoprotectants.

77. SUBJECT AREA: How Cells Are Studied
QUESTION: Disadvantages of transmission electron microscopy over light microscopy include all the following EXCEPT:

A: it can not be used to view living samples.
B: its electron beam damages the sample.
C: it requires more complex sample preparation.
D: it requires more expensive equipment.
E: its resolution is not as good.

Learning Response E: Correct. In spite of the expense, complexity of sample preparation, radiation damage to the sample, and the inability to use living cells, the 100-fold better resolution using electron microscopy makes it a valuable tool for studying cells.

---

78. SUBJECT AREA: How Cells Are Studied
QUESTION: Metal coating can be used in all of the following techniques EXCEPT:

A: freeze-etch microscopy
B: freeze-fracture microscopy
C: phase-contrast microscopy
D: scanning electron microscopy
E: transmission electron microscopy

Learning Response C: Correct. Heavy metal coating is used to increase contrast in transmission microscopy and to both highlight three dimensional structure and transmit electrons from the sample in scanning microscopy.

---

79. SUBJECT AREA: How Cells Are Studied
QUESTION: X-ray diffraction:

A: allows determination of three dimensional structure.
B: allows simple interpretation of the resulting pattern.
C: includes the use of transmission electron

microscopy.
D: is limited to protein samples.
E: uses a sample dissolved in water.

Learning Response A: Correct. Using a highly ordered crystal of any molecule, X-ray diffraction produces a complex pattern of signals which allow the three dimensional structure of the molecule in the crystal to be deduced.

---

80. SUBJECT AREA: How Cells Are Studied
QUESTION: All the following can be used to measure changes in cellular ions EXCEPT:

A: a reference electrode.
B: detach patch measurements.
C: fluorescent indicators.
D: ion selective electrodes.
E: selectively permeable materials.

Learning Response B: Correct. Changes in ion concentration are usually measured by comparing signals from an electrode that has been made ion selective through the use of selectively permeable materials to a signal from a reference electrode. Changes can also be measured using indicators which fluoresce when bound to ions. A detached patch enables the flow of ions through a detached membrane segment.

---

81. SUBJECT AREA: How Cells Are Studied
QUESTION: In order to get impermeable molecules into cells, one can use any of the following techniques EXCEPT:

A: electroporation.
B: microinjection.
C: the use of detergent to make a membrane more permeable.
D: the use of membrane-bound vesicles.
E: all of the above are used.

Learning Response E: Correct. While each technique has its disadvantages particularly due to the unwanted changes induced in the cell, all of the above are used in certain instances.

---

82. SUBJECT AREA: How Cells Are Studied
QUESTION: All of the following are true of a fluorescent activated cell sorter EXCEPT:

A: it allows perfect separation of cells.
B: it can sort cells rapidly.
C: it uses die-coupled antibodies.
D: it uses a laser to induce fluorescence.
E: it uses a strong electric field to deflect cells.

Learning Response A: Correct. This technique uses a laser to activate fluorescent dies attached to the cells by antibodies. This allows the rapid sorting of these cells by taking advantage of the effects of a strong electric field on fluorescent and nonfluorescent cells. While producing pure populations of cells, most cells are not sorted.

---

83. SUBJECT AREA: How Cells Are Studied
QUESTION: All of the following are characteristics of eucaryotic cell lines EXCEPT:

A: they are a convenient source of homogeneous cell population.
B: they are identical to the cells from which they are derived.
C: they can be propagated indefinitely.
D: they can be stored in liquid nitrogen for an indefinite period of time.
E: they often grow best when attached to a solid surface.

Learning Response B: Correct. Cell lines have the advantages of being able to be propagated or stored infinitely and produce a uniform population of cells. However, this population is not identical to the parent cells from which the culture was derived.

---

84. SUBJECT AREA: How Cells Are Studied
QUESTION: All of the following are true of selective cleavage of proteins EXCEPT:

A: can be used to distinguish mutant proteins from their normal counterparts.
B: enable the sequencing of the protein.
C: produces a characteristic pattern of products.
D: produces products which can be separated chromatographically or electrophoretically.
E: relies solely on site specific enzymes.

Learning Response E: Correct. This technique relies on enzymatic or chemical means to produce site specific cleavage of proteins into peptides. The resulting peptides can be analyzed by several different techniques to determine their sequence and characteristics.

---

85. SUBJECT AREA: How Cells Are Studied
QUESTION: All of the following are used for isoelectric focusing EXCEPT:

A: a polyacrylamide gel.
B: a pH gradient.
C: an electric charge.
D: SDS.
E: urea.

Learning Response D: Correct. Isoelectric focusing relies on the charged side groups of a protein to cause it to migrate to a specific position in a pH gradient. This is done by applying a charge to a polyacrylamide gel in the presence of urea and a nonionic detergent. It is often followed by standard SDS-PAGE.

---

86. SUBJECT AREA: How Cells Are Studied
QUESTION: Proteins are often fractionated by all of the following, EXCEPT:

A: column chromatography.
B: HPLC.
C: preparative ultracentrifugation.
D: SDS-PAGE.
E: thin-layer chromatography.

Learning Response E: Correct. Electrophoresis, centrifugation, and chromatography are convenient methods for fractionating proteins. Of the chromatographic techniques, thin-layer is only used to resolve small molecules, not proteins.

---

87. SUBJECT AREA: How Cells Are Studied
QUESTION: The sedimentation of a molecule always depends on all of the following EXCEPT:

A: establishment of a gradient.
B: nature of the solution being spun.
C: shape of the molecule.
D: size of the molecule.
E: speed of centrifugation.

Learning Response A: Correct. While important for density gradient centrifugation, the establishment of a gradient plays no role in certain applications, such as the fractionation of cellular organelles, ribosomes, and other large molecules.

---

88. SUBJECT AREA: How Cells Are Studied
QUESTION: Gel-filtration chromatography of proteins relies on:

A: a difference in the size of proteins.
B: differing affinity of the proteins being separated for the matrix.
C: exclusion of small proteins from the resin.
D: ionic interactions with the resin.
E: the application of high pressure.

Learning Response A: Correct. Gel-filtration uses a resin with holes of a defined size. This resin will exclude large proteins, while holding back smaller oncs. It can be performed through either a pressure or gravity driven flow.

---

89. SUBJECT AREA: How Cells Are Studied
QUESTION: Pulse-chase experiments:

A: are used to uniformly label cells.
B: is used to follow a cellular process through time.
C: must use a radionuclide with an extremely short half-life.
D: rely on radioactive ATP.
E: use a scintillation counter for detection of signal.

Learning Response B: Correct. These experiments use a radioactive substrate (for any reaction) for a brief period of time, and follow this with a large amount of non-labeled substrate. This enables the product formed during the brief labeling period to be followed through the cell, and to be distinguished from the newly synthesized products. The cells are usually fractionated in some way, and the products identified through autoradiography or direct scintillation counting.

---

90. SUBJECT AREA: How Cells Are Studied
QUESTION: Monoclonal antibodies:

A: are isolated from the serum of immunized animals.
B: are produced by immunizing a culture of B cells.
C: can only be produced for a limited period of time.
D: represent a mixture of antibodies to various protein sites.
E: require selection to identify appropriate clones.

Learning Response E: Correct. The process of developing monoclonal antibodies involves the removal of B lymphocytes from immunized mice. These are then fused with a B cell tumor line, and the resulting hybridomas grown in selective medium and screened for clones producing the desired antibody. These antibodies are unique to a specific site on a protein, and can be produced indefinitely by the cell line.

91. SUBJECT AREA: How Cells Are Studied
QUESTION: Northern blotting involves all of the following EXCEPT:

A: detection of RNA.
B: digestion of DNA with restriction nucleases.
C: incubation with a sequence specific probe.
D: techniques similar to Southern blotting.
E: transfer of nucleic acid to paper.

Learning Response B: Correct. Southern blotting is used to detect specific RNAs. It is performed by the electrophoresis of the RNA through a gel, the transfer of the separated RNA to nitrocellulose, and the detection of the message of interest by a specific probe. The technique is analogous to that used in Southern blotting to detect DNA restriction fragments.

92. SUBJECT AREA: How Cells Are Studied
QUESTION: DNA sequencing requires all of the following EXCEPT:

A: a single template clone.
B: products with one defined end.
C: radioactive, fluorescent or visible light detection.
D: resolution of products by size.
E: termination or cleavage reactions specific to each nucleotide.

Learning Response A: Correct. There are now several techniques which can be used to sequence DNA. All of them require a product to be produced with one defined end and one end depending on a nucleotide specific termination or cleavage. These products are resolved by size and imaged through detection of radioactive or fluorescent labels associated with them. The specificity of primer hybridization will permit a single template mixture to be used in several different primer reactions.

93. SUBJECT AREA: How Cells Are Studied
QUESTION: The following are true of restriction endonucleases

EXCEPT:

A: they are an interesting byproduct of recombinant DNA technology.
B: they can be used to evaluate the degree to which sequences are related.
C: they can be used to map a piece of DNA.
D: they cut at specific DNA sequences.
E: they often leave a distinctive overhang.

Learning Response A: Correct. Restriction enzymes perform site specific cleavage in DNA. This not only enables sequences to be mapped and compared, but the ability to specifically cut and rejoin DNA was critical to the development of recombinant DNA techniques.

---

94. SUBJECT AREA: How Cells Are Studied
QUESTION: DNA footprinting:

A: is performed on single stranded DNA.
B: is used to probe for DNA-DNA interactions.
C: produces a characteristic footprint of bands.
D: uses a modification of the chemical sequencing method.
E: uses a region of DNA labeled on both ends.

Learning Response D: Correct. This technique uses a region of double stranded DNA which is labeled on only one end. It is incubated with protein, and nucleotide-specific cleavage of the DNA is prevented by the presence of bound protein. This produces a gap in the ladder of bands which is referred to as a footprint.

---

95. SUBJECT AREA: How Cells Are Studied
QUESTION: The specificity of hybridization allows for all of the following EXCEPT:

A: analysis of tissue specific expression.
B: chemical sequencing of DNA.
C: identification of a specific DNA sequence.

D: identification of related genes.
E: the detection of mutations.

Learning Response B: Correct. The high specificity of nucleotide interactions can be used to detect specific sequences either for the purposes of their identification or to distinguish them from slightly different or mutated sequences. At a lower stringency it can be used to detect related but not identical sequences. This property is also used for enzymatic sequencings and *in situ* hybridization which allows expression to be localized to single cells.

---

96. SUBJECT AREA: How Cells Are Studied
QUESTION: The following are advantages to the use of recombinant technology in studying proteins EXCEPT:

A: by determining the sequence it provides insight into the protein structure and function.
B: it allows production of large amounts of minor proteins.
C: it enables specific changes to be made in the protein to analyze their effects.
D: it enables the expression of a protein to be eliminated from a cell.
E: it is free from the ethical considerations that play a role in other biological studies.

Learning Response E: Correct. This technology allows many diverse aspects of protein function to be studied. However, there are ethical considerations raised by the manipulation of an organism's genetic material.

---

97. SUBJECT AREA: Basic Genetic Mechanisms
QUESTION: The polymerase chain reaction (PCR) requires all of the following EXCEPT:

A: a double stranded DNA template.
B: a temperature stable polymerase.
C: cycles of at least two temperatures.
D: two primers in opposite orientation.

E: all of the above are required.

Learning Response A: Correct. While double stranded DNA is the typical template for these reactions, a single stranded template, such as the product of reverse transcriptase, can be used. The first round of PCR extension will produce a double stranded template, and the reaction proceeds normally from this point.

---

98. SUBJECT AREA: Basic Genetic Mechanisms
QUESTION: Which of the following is true concerning transgenic animals:

A: foreign DNA must be introduced by microinjection.
B: the single copy of a transgene in transgenic flies inserts at a specific site in the genome.
C: transgenic animals may be tested for expression, but do not pass the gene to offspring.
D: transgenic mice are produced by injecting into the female pronucleus at the single cell stage.
E: transgenic mice may contain single or multiple copies of the gene of interest.

Learning Response E: Correct. While there are procedures to introduce DNA by transfection into stem cells in mice, microinjection into the male pronucleus is more common. This is similar to the process for flies, but single or multiple copies may insert, as opposed to the single randomly integrating copy usually seen in flies. This gene, once inserted into the genome, is inherited by the progeny.

---

99. SUBJECT AREA: Basic Genetic Mechanisms
QUESTION: Which of the following techniques are used in chromosome walking?

I. cDNA library preparation.
II. colony or plaque purification (isolation).
III. Southern hybridization.

A: I only.
B: II only.

C: III only.
D: I and II.
E: II and III.

Learning Response E: Correct. To perform chromosome walking, requires the preparation of at least one genomic library, and combines the hybridization of probe and purification of either colonies or plaques (λ plaques are more commonly used than plasmid colonies for this technique) with Southern hybridization of genomic DNA to confirm linkage.

---

100. SUBJECT AREA: Basic Genetic Mechanisms
QUESTION: Which of the following is true about cDNA libraries:

A: each clone is represented equally in the library.
B: they are composed of mRNA molecules placed into a plasmid or viral vector.
C: they are the same from every tissue.
D: they can be used to identify promoter sequences.
E: none of the above is true.

Learning Response E: Correct. cDNA libraries are constructed from the products of a reverse transcription of cellular mRNA (cDNA), placed in a cloning vector. The relative copy number is reflective of the levels in the tissue used, which differ from tissue to tissue, and within the same tissue at different times in development.

---

101. SUBJECT AREA: Basic Genetic Mechanisms
QUESTION: Which of the following statements are true of all viruses:

I. they carry sequences coding for proteins important for viral integration or replication.
II. they have a capsid structure composed of protein for their outer coat.
III. they liberate themselves from infected cells by lysing them.
IV. their genome can be single or double stranded, and composed of DNA or RNA.

A: I and III are true.
B: II and IV are true.
C: II and III are true.
D: I and IV are true.
E: all are true.

Learning Response D: Correct. While some viruses form a capsid structure, and liberate themselves by lysing their host, others bud from their host, and take with them an outer coat of host membrane.

---

102. SUBJECT AREA: Basic Genetic Mechanisms
QUESTION: All tumor viruses:

A: are rendered defective by the transforming genes they carry.
B: carry sequences for homologues of cellular proteins important in cell growth control.
C: integrate into the genome.
D: require reverse transcriptase.
E: all of the above are true.

Learning Response C: Correct. While all of the above are true for RNA tumor viruses, they do not apply to DNA tumor viruses, except for the requirement of integration for transformation.

---

103. SUBJECT AREA: Basic Genetic Mechanisms
QUESTION: RFLP analysis:

A: relies on genetic recombination to locate and isolate a gene of interest.
B: relies on the different restriction fragment lengths produced by different restriction enzymes.
C: requires enough knowledge of the gene of interest to produce a probe specific to it.
D: requires knowledge of the chromosome to which the probe sequence hybridizes.
E: uses northern hybridization techniques to detect the DNA bands.

Learning Response A: Correct. Through the use of Southern hybridization, this technique takes advantage of the different patterns of restriction fragments produced by a single enzyme, based on allelic differences, and the relative rate of recombination between sequences separated by different distances, to locate genes of interest, first to a chromosome, and then to a specific region of the chromosome, eventually allowing its cloning.

---

104. SUBJECT AREA: Basic Genetic Mechanisms
QUESTION: A ribosome:

A: carries out mRNA processing.
B: contains a single RNA molecule but over fifty proteins.
C: changes catalytic activity when encountering a release factor in its A-site.
D: transfers the growing peptide from the tRNA occupying its A-site to that in its P-site.
E: moves along the mRNA from 3' to 5'.

Learning Response C: Correct. When a release factor binds a termination codon in the A-site of a ribosome, rather than catalyzing the transfer of the growing peptide from the P-site tRNA to that in the A-site, and moving along the mRNA from 5' to 3', it instead transfers the protein to a water molecule, and dissociates from the mRNA.

---

105. SUBJECT AREA: Basic Genetic Mechanisms
QUESTION: Which of the following inhibitors, which releases the nascent peptide, blocks protein synthesis in both eucaryotes and procaryotes?

A: anisomycin.
B: erythromycin.
C: puromycin.
D: rifamycin.
E: streptomycin.

Learning Response C: Correct. Puromycin acts in both eucaryotes and procaryotes by transferring itself to the growing nascent chain, causing the release of the chain, and terminating the synthesis of the protein.

106. SUBJECT AREA: Basic Genetic Mechanisms
QUESTION: Which of the following is true of transcription?

A: it is mediated by the enzyme reverse transcriptase.
B: it proceeds along the template strand from 5' to 3'.
C: it produces different products, dependent on the orientation of the polymerase.
D: it terminates at the end of a DNA strand.
E: its product is called mRNA.

Learning Response C: Correct. Proceeding along the template strand from 3' to 5', RNA polymerase produces a product (mRNA, tRNA, or rRNA). This action both begins and terminates at specific sequences of the template DNA.

---

107. SUBJECT AREA: Basic Genetic Mechanisms
QUESTION: Which of the following statements about tRNA is NOT true?

A: different tRNAs have similar structures.
B: it contains several modified nucleotides.
C: it is linked to amino acids by specific tRNA synthetases.
D: it relies on intermolecular base pairing to produce its characteristic structure.
E: all of the above are true.

Learning Response D: Correct. The classical stem-looping structure of tRNA is the result of intra-strand base pairing.

---

108. SUBJECT AREA: Basic Genetic Mechanisms
QUESTION: Which of the following account for the degeneracy of the genetic code?
I. multiple codons for a single tRNA.
II. multiple tRNA synthetases for a single tRNA.
III. multiple tRNAs for single amino acids.

A: I and II.

B: II and III.
C: I and III.
D: I, II, and III.
E: none of the above.

Learning Response C: Correct. While both I and III are in part responsible for the codon degeneracy seen, II would result in a different type of degeneracy, with a single codon resulting in multiple possible amino acids, a type of degeneracy which does not occur.

---

109. SUBJECT AREA: Basic Genetic Mechanisms
QUESTION: Which inhibitor of protein synthesis has a similar effect on procaryotes as anisomycin has on eucaryotes (blocking of translocation)?

A: chloramphenicol
B: erythromycin
C: puromycin
D: streptomycin
E: tetracycline

Learning Response B: Correct. In procaryotes, erythromycin blocks the translocation of the nascent peptide-tRNA complex from the A-site to the vacant P-site of a ribosome, the action of anisomycin in eucaryotes.

---

110. SUBJECT AREA: Basic Genetic Mechanisms
QUESTION: Which of the following are characteristics of eucaryotic translation?
I. coding sequences for multiple proteins are contained on a single mRNA.
II. multiple ribosomes can translate a single mRNA at once.
III. the mRNA usually has a 7-mG cap at its 5' end.

A: I and II.
B: II and III.
C: I and III.
D: I, II and III.

E: none of the above.

Learning Response B: Correct. Eucaryotic mRNA, with extremely rare exceptions, contains sequence coding for only a single protein.

---

111. SUBJECT AREA: Basic Genetic Mechanisms
QUESTION: Which of the following enzymes is NOT directly involved in excision repair of damaged DNA?

A: DNA ligase
B: DNA methylase
C: DNA polymerase
D: DNA repair nuclease
E: All of the above are involved

Learning Response B: Correct. DNA excision repair involves the recognition and removal of damaged sequence by repair nucleases, the filling of the gap thus created by polymerase, and finally, the sealing of the newly synthesized region of the formerly damaged strand to the remaining correct portion of that strand by ligase.

---

112. SUBJECT AREA: Basic Genetic Mechanisms
QUESTION: The rate of mutation of a particular DNA sequence depends on all of the following EXCEPT:

A: the ability of the cell to repair damage.
B: the ability of the protein made from the sequence to tolerate changes in amino acid sequence.
C: the amount of exposure of the DNA to ultraviolet light and chemicals.
D: the evolutionary distance between the sequence and that in a related species.
E: all of the above are true.

Learning Response D: Correct. The evolutionary distance will be reflected by the difference between the two sequences, but this is a result of the mutation rate, rather than the mutation rate depending on it.

113. SUBJECT AREA: Basic Genetic Mechanisms
QUESTION: Which of the following occur during SOS repair in *E. coli*?

A: production of a transcriptional repressor to block production of normal repair enzymes.
B: an increase in the mutation rate.
C: a decrease in cell survival.
D: the production of recBCD.
E: none of the above.

Learning Response B: Correct. The induction of the SOS repair involves the production of recA, which destroys a repressor of the genes of proteins involved in repair, resulting in an increased survival time, but also an increased mutation rate (over the rate seen in undamaged cells).

---

114. SUBJECT AREA: Basic Genetic Mechanisms
QUESTION: The short (1000-2000 bp) Okazaki fragments in bacteria:

A: are produced by the primase.
B: are synthesized in the 3' to 5' direction.
C: exist only transiently during replication.
D: result from leading strand synthesis.
E: result from prematurely terminated replication.

Learning Response C: Correct. Okazaki fragments are the product of lagging strand synthesis by DNA polymerase, and exist only until joined to the remaining lagging strand by ligase.

---

115. SUBJECT AREA: Basic Genetic Mechanisms
QUESTION: Proteins involved in DNA replication include all of the following except:

A: helicase.
B: helix destabilizing proteins.
C: primase.

D: topoisomerase.
E: all of the above are involved.

Learning Response E: Correct. DNA replication requires, in addition to DNA polymerase, a primase to initial strand synthesis, a helicase to unwind the DNA ahead of the replication complex and topoisomerase to release the strain in the DNA caused by this unwinding, and helix destabilizing proteins to keep the DNA from reannealing or forming stem-loop structures before completion of lagging strand synthesis.

---

116. SUBJECT AREA: Basic Genetic Mechanisms
QUESTION: Which of the following is true of DNA methylation, as it relates to mismatch repair?

A: methylation distinguishes the parent strand, so the new strand can be repaired.
B: methylation occurs immediately following replication.
C: mismatched bases are methylated to identify them for removal.
D: only mismatches in the GATC methylation sequence can be repaired .
E: only when both strands of the DNA are methylated can repair proceed.

Learning Response A: Correct. For a short period of time after replication, only the parent strand is methylated, allowing the repair mechanism to distinguish between the two strands. It then repairs any mismatches by altering the daughter (unmethylated) strand.

---

117. SUBJECT AREA: Basic Genetic Mechanisms
QUESTION: Which of the following is NOT true of topoisomerase I?

A: it breaks both strands of the DNA, resulting in free rotation.
B: it cannot separate interlocked rings of DNA.
C: it releases rotational strain caused by the unwinding of the replication fork by helicase.

D: it transiently covalently links itself to the DNA.
E: all of the above are true of topoisomerase I.

Learning Response A: Correct. Topoisomerase covalently links itself to one strand of the DNA, causing a single stranded break and allowing free rotation to reduce the rotational strain, and before rejoining the DNA strand, and removing itself from covalent association.

---

118. SUBJECT AREA: Basic Genetic Mechanisms
QUESTION: The following protein is only involved in DNA recombination:

A: DNA ligase.
B: recA.
C: recBCD.
D: SSB.
E: all of the above are involved in this process.

Learning Response C: Correct. While recA is also involved in the SOS repair response, and SSB and ligase also play roles in replication, the actions of recBCD appears to be limited to recombination.

---

119. SUBJECT AREA: Basic Genetic Mechanisms
QUESTION: Which of the following recombinatorial events is thought to involve synthesis?

A: formation of a cross strand exchange.
B: gene conversion.
C: insertion of λ phage into the genome.
D: protein directed branch migration.
E: none of the above involve DNA synthesis.

Learning Response B: Correct. Gene conversion is thought to involve a strand displacement by DNA polymerase, with the displaced strand in turn replacing its homologue on another chromosome (or another copy of the gene on the same chromosome, as in hemoglobin α), the degradation of the replaced strand and ligation of nicks.

120. SUBJECT AREA: Basic Genetic Mechanisms
QUESTION: All of the following activities of recA are important to recombination EXCEPT:

A: binding of both single and double stranded DNA.
B: cleavage of the SOS repressor.
C: cooperation with SSB.
D: hydrolysis of ATP.
E: protein directed strand displacement.

Learning Response B: Correct. All of these activities of recA are critical to its activity in recombination except for its cleaving of the SOS repressor, which while important to DNA damage repair, is not involved in the recombination process.

---

121. SUBJECT AREA: The Plasma Membrane
QUESTION: Receptor mediated endocytosis of LDL by the LDL receptor:

A: blocks mannose phosphate recognition.
B: upregulates the rate of cholesterol synthesis.
C: is mediated by cholesterol coated pits.
D: leads to foam cell formation.
E: requires receptor clustering.

Learning Response E: Correct. Cellular cholesterol is increased by the binding of LDL to receptors, which are then clustered into clathrin coated pits, and endocytosed.

---

122. SUBJECT AREA: The Plasma Membrane
QUESTION: All the following characterize phospholipids EXCEPT:

A: can form micelles.
B: form a lipid monolayer.
C: have a polar head group.
D: have two hydrocarbon tails.

E: are amphipathic.

Learning Response B: Correct. A typical phospholipid has an amphipathic nature due to a polar head group and two hydrophobic tails. In water it can form either a micelle or bilayer.

---

123. SUBJECT AREA: The Plasma Membrane
QUESTION: Cholesterol:

A: exists in trace amounts in eucaryotic cells.
B: has a similar structure to phospholipids.
C: is confined to the outer leaflet of the bilayer.
D: serves to prevent hydrocarbon chains from crystallizing.
E: tends to make a membrane more fluid.

Learning Response D: Correct. Cholesterol is made up of a planar steroid ring with a short hydrocarbon tail and a small polar head group. Eucaryotic plasma membranes contain large amounts of cholesterol which can readily distribute between the bilayer leaflets and serve both to decrease fluidity and prevent crystallization of the membrane.

---

124. SUBJECT AREA: The Plasma Membrane
QUESTION: Studies of liposomes and black membranes revealed the following movement of membrane phospholipids EXCEPT:

A: flexing of hydrophobic arms.
B: lateral diffusion in the membrane.
C: rapid exchange from one surface to the other.
D: rapid rotation along the long axis.
E: all of the above were observed.

Learning Response C: Correct. The motion within a lipid monolayer is extremely fluid with rotational lateral and flexive movement all occurring freely. However, the flip-flop of phospholipids across the bilayer rarely occurs.

---

125. SUBJECT AREA: The Plasma Membrane
QUESTION: Which of the following statements concerning the plasma membrane is true?

A: glycolipids are rarely found.
B: it serves as a solvent for transmembrane proteins.
C: lipids are distributed symmetrically.
D: only contains glycerol based lipids.
E: tends to be more charged on the extracellular leaflet.

Learning Response B: Correct. Plasma membrane is made up of phospholipids, glycolipids and cholesterol of which only the latter is evenly distributed. This uneven distribution creates a membrane with glycolipids facing the extracellular space while the cytoplasmic face is charged. The variability in the membrane provides a special environment for transmembrane proteins.

---

126. SUBJECT AREA: The Plasma Membrane
QUESTION: Which of the following statements about membrane associated proteins is true?

A: they all interact directly with the membrane.
B: they are hydrophobic in nature.
C: they can be linked covalently to membrane lipids.
D: they can only be released by disrupting the bilayer.
E: they must pass through the membrane at least once.

Learning Response C: Correct. This class of proteins includes transmembrane proteins and proteins which interact with the lipid bilayer through a covalent attachment. It also includes proteins whose association with the membrane is due to noncovalent interactions with other membrane proteins and can be dissociated without disrupting the bilayer. While they must have or be associated with a hydrophobic protein domain or lipid, they can also contain large hydrophilic regions.

---

127. SUBJECT AREA: The Plasma Membrane

QUESTION: When proteins in membranes are exposed to SDS, all the following happen EXCEPT:

A: lipid detergent micelles are formed.
B: the bilayer is destroyed.
C: the protein becomes irreversibly denatured.
D: the protein becomes inactive.
E: the protein is solubilized.

Learning Response C: Correct. By interacting with the hydrophobic surfaces of the protein SDS will solubilize, denature and inactivate the protein. However, in some cases, removal of the detergent allows activity to be restored.

---

128. SUBJECT AREA: The Plasma Membrane
QUESTION: The advantages to using red blood cells to study the plasma membrane include all the following EXCEPT:

A: they can be easily lysed by a hypertonic salt solution.
B: they can be resealed in either orientation.
C: they contain no internal membranes
D: they represent a relatively pure population of cells.
E: red blood cells are readily available.

Learning Response A: Correct. Red blood cells are easily obtainable and represent a pure population with only plasma membrane. Lysis by hypotonic salt produces ghosts which can be sealed in either orientation and contain test solutions.

---

129. SUBJECT AREA: The Plasma Membrane
QUESTION: Freeze-fracture of red blood cells:

A: occurs along the surface of the membrane.
B: produces a cytoplasmic E face.
C: produces an exposed face which is directly observed.
D: reveals primarily band 3 molecules.
E: reveals proteins which are evenly distributed on the two faces.

Learning Response D: Correct. Freeze-fracture splits the cell membrane between the lipid layers producing an external E face and a cytoplasmic P face. A platinum replica is then made and observed. In red blood cells this reveals the band 3 protein predominantly on the P face.

---

130. SUBJECT AREA: The Plasma Membrane
QUESTION: All the following are true of membrane proteins EXCEPT:

A: migration ceases following bleaching.
B: their mobility is proven by patching.
C: they are sometimes confined to specific membrane domains.
D: they can be followed using fluorescent antibodies.
E: they can defuse in any plane of the membrane.

Learning Response A: Correct. The free mobility of proteins within a membrane can be studied with fluorescent antibodies. These enable the observation of patching and the diffusion mediated recovery from bleaching. They also enable the study of irregular distribution of membrane proteins into specific cellular domains.

---

131. SUBJECT AREA: The Plasma Membrane
QUESTION: Carbohydrate in the plasma membrane:

A: is confined to the cytoplasm.
B: is linked to membrane proteins through a single distinct chemical bond.
C: is referred to as lectin.
D: is symmetrically distributed in the membrane.
E: plays an important role in cell-cell recognition.

Learning Response E: Correct. Carbohydrates are linked to membrane proteins on the non-cytosolic face through one of two distinctive linkages. They are bound by lectin and are involved in the interaction of the cell with its surroundings.

132. SUBJECT AREA: The Plasma Membrane
QUESTION: Ions that cross a membrane:

A: always move into a cell.
B: are enabled to pass through channels by a conformational change.
C: diffuse freely across the membrane.
D: do not require energy for facilitated diffusion.
E: require energy for transport with a gradient.

Learning Response D: Correct. Only uncharged molecules can cross a membrane by passive diffusion. Ions pass with a gradient through a channel in a process called facilitated diffusion, or by a carrier protein in a process involving a conformational change. Energy is only required for transport against a gradient.

---

133. SUBJECT AREA: The Plasma Membrane
QUESTION: All of the following are true of membrane transport proteins, EXCEPT:

A: they are multipass transmembrane proteins.
B: they convey ions in only one direction.
C: they require energy for motion against a gradient.
D: they show similar kinetics to enzymes.
E: they undergo a conformational change.

Learning Response B: Correct. Membrane transport proteins use a conformational change to transport ions, which requires energy to transport against the gradient. These multipass molecules are similar to enzymes in their kinetics, and can convey one or several ions in the same or in opposing directions.

---

134. SUBJECT AREA: The Plasma Membrane
QUESTION: The sodium potassium pump:

A: is incapable of producing ATP.
B: controls the amount of water in a cell.

C: exchanges three potassiums for two sodiums.
D: is stimulated by ouabain.
E: works by passive diffusion.

Learning Response B: Correct. By controlling the osmotic balance of a cell, this pump causes water to move by passive diffusion, however its main function is the ATP control exchange of three sodiums for two potassiums, a reaction inhibited by ouabain. By adjusting ion concentrations, the exchange can be made to occur in reverse producing ATP.

---

135. SUBJECT AREA: The Plasma Membrane
QUESTION: Active transport:

A: can work with a gradient to produce ATP.
B: is used to establish gradients.
C: requires hydrolysis of ATP.
D: uses a counter ion to balance the charge.
E: will only transport ions.

Learning Response B: Correct. Active transport can occur either through the hydrolysis of ATP or by using as an energy source a gradient that has already been established through active transport. This process sometimes involves a counter ion and can also be used to transport other molecules such as sugars.

---

136. SUBJECT AREA: The Plasma Membrane
QUESTION: Neuromuscular transmission involves all the following EXCEPT:

A: ligand-gated cation channel
B: voltage-gated calcium channel.
C: voltage-gated potassium channel.
D: voltage-gated sodium channel.
E: all the above are involved.

Learning Response C: Correct. This process involves an action potential opening the voltage-gated calcium channel at the nerve terminal. This induces neurotransmitter release which in turn opens a cation channel to sodium. This sodium in turn opens a voltage-gated sodium channel in the plasma membrane and leads to release of calcium from the sarcoplasmic reticulum.

---

137. SUBJECT AREA: The Plasma Membrane

QUESTION: All the following are properties of ionophores EXCEPT:

A: they are man made.
B: there are two classes of ionophores.
C: they are selective for specific ions.
D: they dissolve in lipid bilayers.
E: they increase membrane permeability.

Learning Response A: Correct. Ionophores were originally isolated from microorganisms. They dissolve in membrane and form either channels or serve as mobile carriers for specific ions.

---

138. SUBJECT AREA: The Plasma Membrane

QUESTION: All the following accurately describe patch-clamp recording EXCEPT:

A: can be used to study cells too small for other techniques.
B: can measure the effect of a single channel.
C: can record from ion channels in cells that are not electrically excitable.
D: involves the use of an intercellular electrode.
E: reveal an all-or-nothing opening of channels.

Learning Response D: Correct. This technique involves the measurements of channel activity in a small region of membrane that could contain only a single channel, even cells too small to measure with intercellular electrodes. It has revealed that differences in current are due solely to the number of channels open and not to a gradation in the extent of flow through a single channel.

---

139. SUBJECT AREA: The Plasma Membrane

QUESTION: Which of the following accurately describes exocytosis?

A: it involves fusion of two bilayers.
B: it involves the release of secretory vesicles from the constitutive pathway.
C: it is used to bring molecules into the cell.
D: it only occurs in secretory cells.
E: only the regulated pathway proceeds from the Golgi apparatus.

Learning Response A: Correct. Exocytosis involves the release of material through the cell membrane. There are two pathways for this: the regulative pathway which uses secretory vesicles and the constitutive pathway. These diverge at the Golgi apparatus. This process also carries membrane proteins to the plasma membrane of all cells.

---

140. SUBJECT AREA: The Plasma Membrane
QUESTION: All the following refer to coated vesicles EXCEPT:

A: they are not all clathrin coated.
B: they are the product of phagocytosis.
C: they develop from coated pits.
D: they have a highly ordered coat structure.
E: they result from ingestion of fluids and solids.

Learning Response B: Correct. Pinocytosis involves the ingestion of fluids and solutes into coated pits which then pinch off as coated vesicles. These have highly structured coats which in one form use clathrin to provide that structure.

---

141. SUBJECT AREA: The Plasma Membrane
QUESTION: All of the following are characteristics of receptor-mediated endocytosis EXCEPT:

A: it increases the efficiency of absorption by three orders of magnitude.
B: it involves specific molecular interactions.

C: it is also referred to as fluid-phase endocytosis.
D: it occurs in clathrin coated pits and vesicles.
E: macromolecules enter cells as receptor-ligand complexes.

Learning Response C: Correct. This procedure involves specific recognition of solutes by receptors and their endocytosis in clathrin coated pits. It is over 1000-fold more efficient than fluid phase endocytosis.

---

142. SUBJECT AREA: The Plasma Membrane
QUESTION: All the following accurately describe phagocytosis EXCEPT:

A: in mammals it involves an antibody based zippering mechanism.
B: in mammals only macrophages act in this way.
C: it is a specialized form of endocytosis.
D: it is a localized response to cell-cell contact.
E: it is promoted by specific receptors in mammals.

Learning Response B: Correct. This specialized form of endocytosis is carried out in mammals by macrophages and neutrophils. It involves receptor recognition of antibody coated target cells. Through a membrane-zippering mechanism, the target cell is surrounded and internalized.

---

143. SUBJECT AREA: The Plasma Membrane
QUESTION: Which of the following is a characteristic of endosomes?

A: they are bypassed in transcytosis.
B: they are the site of protein degradation.
C: they exist as a uniform population.
D: they pass all their contents to lysosomes.
E: they receive vesicles directly from the Golgi apparatus.

Learning Response E: Correct. Endosomes exist as two distinct compartments that receive endocytic vesicles. Their contents are either returned to the membrane from which they came, passed to a different part of the plasma membrane (transcytosis) or passed to the lysosomes. This last fate is shared by lysosomal enzymes received by the endosome directly from the Golgi apparatus.

---

144. SUBJECT AREA: The Plasma Membrane

QUESTION: Endocytosis is thought to be involved in all the following processes EXCEPT:

A: capping.
B: cell locomotion.
C: cell nutrition.
D: transcytosis.
E: all of the above.

Learning Response E: Correct. Endocytosis is not only involved in bringing nutrients into the cell, but is also involved in redirecting cellular membranes and proteins in transcytosis, locomotion and capping.

---

145. SUBJECT AREA: Energy Conversion: Mitochondria and Chloroplasts

QUESTION: Which of the following accurately describes mitochondrial structure?

A: they can be fractionated into four fractions.
B: they contain an outer and inner membrane otherwise known as the matrix.
C: they contain cristae which have the same structure in all cells.
D: they contain intermembrane space surrounded by a lipid bilayer.
E: they maintain a constant shape over time.

Learning Response A: Correct. A mitochondria has two membranes. The inner membrane which separates the matrix from the membrane space demonstrates invaginations termed cristae which have different structures in different cell types. The shape of mitochondria changes rapidly in a living cell.

146. SUBJECT AREA: Energy Conversion: Mitochondria and Chloroplasts
QUESTION: All of the following are aspects of the fatty acid oxidation cycle EXCEPT:

A: it directly produces two molecules of ATP.
B: it feeds into the citric acid cycle.
C: it produces energy during a fast state.
D: it results in production of acetyl CoA.
E: it utilizes triglycerides as its fat source.

Learning Response A: Correct. This cycle which is active in the fasted state uses fatty acids from triglycerides to produce $FADH_2$, NADH, and acetyl CoA which feeds into the citric acid cycle.

---

147. SUBJECT AREA: Energy Conversion: Mitochondria and Chloroplasts
QUESTION: The following are true of the citric acid cycle EXCEPT:

A: it accounts for two thirds of the total oxidation of carbon compounds in cells.
B: it leads to the direct production of ATP.
C: it is also known as the Krebs cycle.
D: two carbons enter with each cycle.

Learning Response B: Correct. This is the only untrue statement. The citric acid cycle (Krebs cycle) uses two carbons from acetyl CoA to produce energy in the form of NADH, $FADH_2$ and GTP as well as two molecules of carbon dioxide. ATP is produced indirectly by oxidative phosphorylation.

---

148. SUBJECT AREA: Energy Conversion: Mitochondria and Chloroplasts
QUESTION: The transfer of electrons to oxygen:

A: directly produces ATP.
B: establishes a proton gradient.
C: is performed by one enzyme.
D: occurs in the mitochondrial matrix.

E: relies on the ion permeability of the inner membrane.

Learning Response B: Correct. The electrons from the citric acid cycle are transferred to oxygen by three enzyme complexes in the mitochondrial membrane. These produce a proton gradient which pass back into the matrix through a proton pump, producing ATP.

---

149. SUBJECT AREA: Energy Conversion: Mitochondria and Chloroplasts
QUESTION: The proton gradient drives which of the following?

I. the exchange of ADP for ATP.
II. the import of phosphates.
III. the import of pyruvates.
IV. the production of ATP.

A: I, II and III.
B: I, II and IV.
C: I, III and IV.
D: II, III and IV.
E: I, II, III and IV.

Learning Response D: Correct. The proton gradient not only drives ATP production, but is also coupled to the production of pyruvate and phosphate. It is the production of ATP that drives the ADP-ATP antiporter.

---

150. SUBJECT AREA: Energy Conversion: Mitochondria and Chloroplasts
QUESTION: Oxidative phosphorylation produces a maximum of:

A: 1 molecule of ATP for each acetyl CoA.
B: 3 molecules of ATP for each $FADH_2$.
C: 12 molecules of ATP for each NADH.
D: 24 molecules of ATP from the acetyl CoA coming from glucose.
E: 36 molecules of ATP for each molecule of palmitate.

Learning Response D: Correct. By producing 3 ATP from each NADH and 2 from each $FADH_2$, oxidative phosphorylation results in 12 molecules of ATP from each acetyl CoA, or 24 from glucose and 96 from palmitate. To this should be added the ATP produced in the formation of the acetyl CoA from those substrates building a total of 36 and 129.

---

151. SUBJECT AREA: Energy Conversion: Mitochondria and Chloroplasts
QUESTION: All of the following are true of the enzyme ATP synthetase EXCEPT:

A: can be purified.
B: can be studied in inside-out submitochondrial particles.
C: can hydrolyze ATP to pump protons.
D: can not be reconstituted into membranes.
E: represents 15% of inner membrane protein.

Learning Response D: Correct. This enzyme has been studied in submitochondrial particles and has been purified and reconstituted into membrane. It can not only produce ATP, using a proton gradient, but can produce the gradient through hydrolysis of ATP.

---

152. SUBJECT AREA: Energy Conversion: Mitochondria and Chloroplasts
QUESTION: The respiratory chain:

A: includes cytochrome P450 enzymes.
B: is localized to the mitochondrial matrix.
C: uses three polypeptides.
D: uses six different iron sulfur centers.
E: utilizes a magnesium atom to accept electrons.

Learning Response D: Correct. The respiratory chain is localized to the inner membrane of the mitochondria. It contains three multiprotein complexes which pass electrons to oxygen through several hemes, iron sulfur centers and copper atoms.

---

153. SUBJECT AREA: Energy Conversion: Mitochondria and Chloroplasts
QUESTION: Proton ionophores will do all of the following EXCEPT:

A: will dissipate the proton gradient.

B: will increase oxygen uptake.
C: will increase PTP synthesis.
D: will remove respiratory control.
E: will uncouple electron transport.

Learning Response C: Correct. By allowing free movement of protons, ionophores will eliminate the proton gradient and with it the feedback regulation on respiration. This will result in increased oxygen used without producing any ATP.

---

154. SUBJECT AREA: Energy Conversion: Mitochondria and Chloroplasts
QUESTION: The members of the respiratory chain:

A: are present in equal amounts within a cell.
B: form a structurally ordered chain of proteins.
C: form a super complex.
D: interact by random collision.
E: occur in equal amounts in all cells.

Learning Response D: Correct. Because they interact by random collision, the members of this chain form no permanent or ordered complex. They exist in different amounts within cells and their ratios are different between cells.

---

155. SUBJECT AREA: Energy Conversion: Mitochondria and Chloroplasts
QUESTION: All of the following chloroplast structures are shared by mitochondria EXCEPT:

A: a DNA molecule.
B: a stroma/matrix.
C: a thylakoid space.
D: an intermembrane space.
E: inner and outer membranes.

Learning Response C: Correct. While they share common structures, including membranes and DNA genome, a chloroplast contains an additional compartment, the thylakoid. This structure surrounded by its own membrane is analogous to mitochondrial cristae, invaginations of the mitochondrial inner membrane.

---

156. SUBJECT AREA: Energy Conversion: Mitochondria and Chloroplasts
QUESTION: The following describe the production of complex carbohydrates from carbon dioxide EXCEPT:

A: it is a process unique to chloroplasts.
B: it is catalyzed by ribulose bisphosphate carboxylase.
C: it is sometimes called the dark reactions.
D: it requires a molecule of ATP for each carbon dioxide.
E: it uses energy from photosynthetic electron-transfer reactions.

Learning Response D: Correct. Carbon fixation uses energy in the form of three ATP and one NADH from the light reaction to fix each molecule of carbon dioxide. This process which is unique to chloroplasts is catalyzed by ribulose bisphosphate carboxylase.

---

157. SUBJECT AREA: Energy Conversion: Mitochondria and Chloroplasts
QUESTION: The chlorophyll molecule:

A: absorbs energy in the form of sunlight.
B: functions in photosynthesis by converting absorbed energy into heat.
C: has a porphyrin ring identical to that in heme.
D: has an iron atom at its active center.
E: lacks the hydrophobic tail of heme.

Learning Response A: Correct. Chlorophyll is a heme like molecule but differs in the exact structure of its porphyrin ring, by having magnesium at its center and by having a long hydrophobic tail. In photosynthesis it functions by absorbing energy from light. It then passes this energy to a neighboring chlorophyll molecule, or to an electron acceptor in the form of an excited electron.

---

158. SUBJECT AREA: Energy Conversion: Mitochondria and Chloroplasts
QUESTION: All the following are true of the photochemical reaction center EXCEPT:

A: contains a special pair of chlorophyll molecules.
B: is of relatively recent origin.
C: part of the chloroplasts photosystem.
D: rapidly conveys electrons to more stable molecules.
E: receives electrons funneled from the antenna complex.

Learning Response B: Correct. A photo system consists of a closely linked antenna complex and photochemical reaction center. The latter receives energy in the form of excited electrons from the antenna complex and through special pairs of chlorophyll molecules, to more stable acceptors. The reaction center is thought to be at least three billion years old.

---

159. SUBJECT AREA: Energy Conversion: Mitochondria and Chloroplasts
QUESTION: Photosynthesis in plants involves all of the following EXCEPT:

A: it directly produces ATP.
B: it passes electrons back and forth across the membrane.
C: it produces a proton gradient.
D: it requires several quanta of light to produce NADPH.
E: it uses two photosystems.

Learning Response A: Correct. In noncyclic photophosphorylation, two photo systems are involved. Photosystem II produces electrons which are passed to photosystem I establishing a proton gradient in the process. Photosystem I uses additional quanta of light to produce highly excited electrons which are then used to form NADPH. The proton gradient is used to produce ATP.

---

160. SUBJECT AREA: Energy Conversion: Mitochondria and Chloroplasts
QUESTION: The proton gradient in chloroplasts:

A: can only be produced by photosystem II.
B: is only produced during the production of NADPH.
C: is oriented opposite that of mitochondria.
D: is produced across the inner membrane.
E: is produced by a cytochrome-like complex.

Learning Response E: Correct. Chloroplasts establish a proton gradient in their thylakoid space by using a complex very similar to the cytochrome b-c1 complex of mitochondria and oriented similarly. This gradient can either be established by photosystem II or by electrons diverted from photosystem I driven NADPH synthesis to ATP synthesis.

---

161. SUBJECT AREA: Energy Conversion: Mitochondria and Chloroplasts
QUESTION: The proposal that mitochondria and chloroplasts evolved from a symbiotic relationship include all the following EXCEPT:

A: each carries out reactions similar to a specific class of bacteria.
B: genes for some of their proteins are contained in the nuclear genome.
C: they contain their own DNA.
D: they have a double membrane unlike other organelles.
E: they have their own complete genetic systems.

Learning Response B: Correct. While many of their genes appear to have been transferred to the nuclear DNA, these organelles still possess characteristics that would be expected of an endosymbiont, including those named above.

---

162. SUBJECT AREA: Energy Conversion: Mitochondria and Chloroplasts
QUESTION: The following were probably involved in the evolution of electron transport chains EXCEPT:

A: a switch from the use of fermentation to produce ATP.
B: the development of a proton pump.
C: the development of an ATP independent proton excretion system.
D: the development of the ability to produce a proton gradient.
E: the same energy sources as are used today.

Learning Response E: Correct. Ancient cells probably developed a proton pump which used the ATP derived from fermentation to eliminate the pH gradient that it produced. It is proposed that cells subsequently developed ATP independent methods for balancing pH which eventually became so efficient that they reversed the gradient allowing ATP synthesis from the pumps. A driving force in this process was probably in the available energy sources.

---

163. SUBJECT AREA: Energy Conversion: Mitochondria and Chloroplasts
QUESTION: All of the following characterize the appearance of atmospheric oxygen EXCEPT:

A: it caused some bacteria to abandon photosynthesis.
B: it is evidenced by extensive geological formations of banded iron.
C: it predates the appearance of life.
D: it required the evolution of protective measures.
E: it was derived from the splitting of water.

Learning Response C: Correct. The appearance of oxygen is not an ancient event by evolutionary standards. Derived from the splitting of water, its appearance can be traced in the geologic record and probably resulted in significant evolutionary changes to the organisms living at the time.

---

164. SUBJECT AREA: Energy Conversion: Mitochondria and Chloroplasts
QUESTION: All the following are characteristics of mitochondria EXCEPT:

A: they are capable of division.
B: they contain genes for all of their proteins.
C: they contain their own DNA.
D: they contain their own genetic system.
E: they have similarities to chloroplasts.

Learning Response B: Correct. Both mitochondria and chloroplasts contain their own DNA and genetic systems that are capable of division. However, some of the genes for their proteins are resident in the nuclear DNA.

---

165. SUBJECT AREA: Energy Conversion: Mitochondria and Chloroplasts
QUESTION: Differences between the plant chloroplast and animal mitochondrial genomes include all of the following EXCEPT:

A: ability to produce rRNA.
B: genetic code used.
C: length of genome.
D: number of proteins encoded.
E: number of tRNAs used.

Learning Response A: Correct. While the chloroplast genome is ten times as large and encodes ten times as many genes, they also differ in more substantial ways. The ribosomal genetic code differs from all others both using less tRNAs and a different pattern of codon usage. Still, they both encode their own rRNA.

---

166. SUBJECT AREA: Energy Conversion: Mitochondria and Chloroplasts
QUESTION: The following are true of mitochondrial genomes EXCEPT:

A: in animals they have been sequenced in their entirety.
B: they are subject to a high rate of mutation in animals.
C: they contain introns in some species.
D: they contain no nonessential genetic material.
E: they produce a transcript which is processed.

Learning Response D: Correct. Mitochondrial DNA of some species has been sequenced and found to be subject to a high rate of mutation. While in animals they are highly refined producing transcripts that are processed into rRNA, tRNA, as well as mRNA, the genomes of other groups contain introns and other "junk" DNA.

---

167. SUBJECT AREA: Energy Conversion: Mitochondria and Chloroplasts
QUESTION: In animals, mitochondrial genes can be distinguished from their nuclear counterparts by all the following EXCEPT:

A: they demonstrate non-Mendelian genetics.
B: they form a circular linkage group with other mitochondrial genes.
C: they have distinct sequences from nuclear counterparts.

D: they use different nucleotide bases to encode their proteins.
E: unlike most nuclear genes they contain no introns.

Learning Response D: Correct. While they use the same nucleotides, the sequence of mitochondrial genes are more closely related to those of bacteria then mammalian genes. They demonstrate a maternal pattern of inheritance and form a circular linkage group.

---

168. SUBJECT AREA: Energy Conversion: Mitochondria and Chloroplasts
QUESTION: The import of proteins into mitochondria:

A: allows them to make their own lipids.
B: is so critical that mutations are always fatal.
C: is uniform in all tissues.
D: occurs at specific sites in the double membrane.
E: occurs through passive diffusion.

Learning Response D: Correct. Mitochondria import most of their protein and lipids through an energy requiring process which occurs at specific contact sites where the inner and outer membranes are joined. Because this process can be tissue specific, a defect in a particular cell type need not be fatal to the whole animal.

---

169. SUBJECT AREA: Intracellular Sorting and the Maintenance of Cellular Compartments
QUESTION: The nuclear envelope:

A: contains pores that transport nucleic acids and proteins.
B: does not have an associated cytoskeleton.
C: is composed of a single membrane.
D: is identical to the endoplasmic reticulum.
E: is the site of protein synthesis.

Learning Response A: Correct. The nuclear envelope is made up of two membrane layers. The inner membrane is associated with a cytoskeletal lamina, while the outer appears to be continuous with the endoplasmic reticulum. For nucleic acids and proteins to pass across the membrane, they must go through pores which traverse the double membrane.

---

170. SUBJECT AREA: Intracellular Sorting and the Maintenance of Cellular Compartments

QUESTION: All the following organelles are found in all eucaryotic cells EXCEPT:

A: chloroplast.
B: endoplasmic reticulum.
C: Golgi apparatus.
D: mitochondrion.
E: peroxisome.

Learning Response A: Correct. Of the above, chloroplasts are limited exclusively to the plants. Each of the others are found in all eucaryotes.

---

171. SUBJECT AREA: Intracellular Sorting and the Maintenance of Cellular Compartments

QUESTION: Protein traffic through the Golgi apparatus:

A: becomes divided into different pathways.
B: can not be studied using electron microscopy.
C: involves all proteins produced by the cell.
D: leads to constitutive secretion.
E: occurs within three minutes of synthesis.

Learning Response A: Correct. The traffic of proteins through the cell can be studied by electron microscopy and pulse-chase labeling. This reveals that only membrane and secreted proteins are produced in the ER and are slowly passed on to the Golgi. Here the constitutive and regulated pathways of secretion diverge.

---

172. SUBJECT AREA: Intracellular Sorting and the Maintenance of Cellular Compartments

QUESTION: Signal peptides:

A: are added to the sorted protein.
B: are formed by the interaction of distant parts of the protein.
C: are important for protein targeting.
D: are present in all proteins.
E: are the only form of sorting signal.

Learning Response C: Correct. Signal peptides, which are distinct from signal patches in that the latter consists of multiple regions of protein brought together by the structure, distinguish which proteins are to be sorted into a particular organelle.

---

173. SUBJECT AREA: Intracellular Sorting and the Maintenance of Cellular Compartments

QUESTION: All of the following are true of cellular organelles, EXCEPT:

A: they are bound by membranes.
B: they are unique to eucaryotic cells.
C: they can all be made *de novo* by cells.
D: they carry out unique processes.
E: they have a topology consistent with their proposed evolution.

Learning Response C: Correct. Organelles are membrane bound structures, present only in eucaryotes, which carry out specialized cellular functions. They have membrane topologies consistent with their proposed origins as plasma membrane invaginations or symbiotic ingestion of bacteria. While most of their proteins are encoded by nuclear genes, they appear to propagate through growth and division of existing organelles, rather than by *de novo* organelle synthesis.

---

174. SUBJECT AREA: Intracellular Sorting and the Maintenance of Cellular Compartments

QUESTION: All of the following relate to the rapid degradation of cytosolic proteins, EXCEPT:

A: it can be altered by post-translational modifications.

B: it can involve cross linking with ubiquitin.
C: it depends on the amino-terminal sequence.
D: it occurs to all cytosolic proteins.
E: it uses the same pathway as that which degrades abnormal proteins.

Learning Response D: Correct. The method by which a subset of cytoplasmic proteins are rapidly degraded is the same as degrades misfolded or otherwise abnormal proteins. It can involve ubiquitination, and is dependent on the amino terminal sequence of the protein, which is often modified following translation of the protein.

---

175. SUBJECT AREA: Intracellular Sorting and the Maintenance of Cellular Compartments
QUESTION: Heat-shock proteins:

A: are only involved in the stress response.
B: are synthesized only in response to heat treatment.
C: occur in three main families.
D: promote protein aggregation.
E: seem to be limited to higher eucaryotes.

Learning Response C: Correct. Heat-shock proteins are usually synthesized in response to heat or other stresses, but the members of some of the three families play roles in other cellular functions. Found in all organisms, these proteins prevent protein aggregates from accumulating.

---

176. SUBJECT AREA: Intracellular Sorting and the Maintenance of Cellular Compartments
QUESTION: The nuclear pore:

A: actively transports proteins out of the nucleus prior to mitosis.
B: does not remove the signal peptide following transport.
C: only transports proteins.
D: pass proteins across the membrane.

E: recognizes an amino-terminal signal peptide.

Learning Response B: Correct. Nuclear pores recognize a signal peptide located is any part of a nuclear protein and actively transport the protein into the nucleus. The signal peptide is not removed, and is required to re-import the proteins after the dispersal of the proteins which occurs during each mitosis. The pore also transports RNA out of the nucleus.

---

177. SUBJECT AREA: Intracellular Sorting and the Maintenance of Cellular Compartments

QUESTION: Which of the following accurately describes the mitochondrial signal peptide?

A: a hydrophobic helix.
B: an amphipathic helix.
C: an internal α helix.
D: the amino-terminal amino acid.
E: twelve carboxyl-terminal amino acids.

Learning Response B: Correct. The mitochondrial signal peptide is an amphipathic α helix formed by the amino-terminal twelve amino acids of imported proteins.

---

178. SUBJECT AREA: Intracellular Sorting and the Maintenance of Cellular Compartments

QUESTION: The translocation of proteins into the intermembrane space of the mitochondria involves all of the following EXCEPT:

A: a single signal peptide.
B: a special site in the membrane.
C: a two step process.
D: removal of signal peptides.
E: unfolding of the protein.

Learning Response A: Correct. In order to enter the intermembrane space, a two step process is used. The protein is first unfolded and passed through contact sites between the inner and outer membranes. The signal peptide directing this is then removed, and a second peptide, also eventually removed, targets the protein for transport across the inner membrane and into the intermembrane space.

---

179. SUBJECT AREA: Intracellular Sorting and the Maintenance of Cellular Compartments

QUESTION: Which of the following are true of peroxisomes:

A: they are uniform in content.
B: they have a double membrane.
C: they possess their own DNA.
D: they produce some of their own proteins.
E: they use molecular oxygen.

Learning Response E: Correct. Peroxisomes have a single membrane and, possessing no DNA, must import all of their proteins. While diverse in the enzymes which they contain, they perform reactions which oxidize molecules using oxygen and peroxide.

---

180. SUBJECT AREA: Intracellular Sorting and the Maintenance of Cellular Compartments

QUESTION: All of the following are true of the rough endoplasmic reticulum, EXCEPT:

A: it can be purified by centrifugation.
B: it can be observed using electron microscopy.
C: it is defined by the binding of ribosomes.
D: it is found in different tissues than smooth endoplasmic reticulum.
E: it is the site of synthesis of secreted proteins.

Learning Response D: Correct. Contiguous with the smooth endoplasmic reticulum, the rough endoplasmic reticulum binds ribosomes which are producing proteins to be secreted or inserted into the plasma membrane. This binding of ribosomes to the rough endoplasmic reticulum allows it to be distinguished by centrifugation and electron microscopy.

181. SUBJECT AREA: Intracellular Sorting and the Maintenance of Cellular Compartments

QUESTION: The process of translocation of a secreted protein through the endoplasmic reticulum membrane usually involves all of the following, EXCEPT:

A: a hydrophobic helix.
B: a transmembrane signal recognition particle.
C: binding of the ribosome to the endoplasmic reticulum membrane.
D: ongoing protein synthesis.
E: removal of the signal peptide.

Learning Response C: Correct. The process of translocation across the endoplasmic reticulum membrane begins with the binding of the soluble signal recognition particle to the hydrophobic signal sequence at the amino-terminus of the growing protein. This allows the binding of the ribosome to receptors on the endoplasmic reticulum membrane, and the protein is passed through the membrane as it is translated. To be secreted, the signal peptide must be removed.

---

182. SUBJECT AREA: Intracellular Sorting and the Maintenance of Cellular Compartments

QUESTION: Which is NOT true of the transmembrane peptides which direct membrane insertion:

A: they are cleaved from the protein following transfer.
B: they can be found anywhere in the protein.
C: they can be predicted from sequence.
D: they determine the topology of a membrane protein.
E: several can exist in a single protein.

Learning Response A: Correct. The insertion of a protein into the membrane depends on the location of one or several transmembrane domains in the protein. These can be predicted by the hydrophobicity of the sequence, and determine the topology of the protein in the membrane. While the signal sequence is often removed, this is not the case with the other signals.

---

183. SUBJECT AREA: Intracellular Sorting and the Maintenance of Cellular Compartments

QUESTION: The folding of proteins in the ER lumen involves all of the following, EXCEPT:

A: a varying amount of time.
B: binding of incorrectly folded proteins.
C: rearrangement of disulfide bonds.
D: reducing agents, such as glutathione.
E: stress-response proteins.

Learning Response D: Correct. Proteins are translocated into the endoplasmic reticulum in an unfolded state. There they bind heat-shock-like proteins, which retain them until they are properly folded. Unlike the cytoplasm, the lumen does not contain reducing agents, but instead rearranges disulfide bonds enzymatically.

---

184. SUBJECT AREA: Intracellular Sorting and the Maintenance of Cellular Compartments

QUESTION: The following are true of membrane lipids in the endoplasmic reticulum, EXCEPT:

A: some can be flipped across the membrane.
B: some can replace the carboxyl-termini of proteins.
C: some are synthesized as precursors for Golgi-produced lipids.
D: some are transferred to other membranes by specific exchange proteins.
E: they are synthesized on the lumenal side of the membrane.

Learning Response E: Correct. The production of lipids by the ER occurs on the cytoplasmic leaflet, from which they can be removed by exchange proteins, or flipped to the lumenal side. From the lumenal leaflet, they can be attached to membrane associate proteins, or passed on to the Golgi apparatus, where some are used for the production of sphingomyelin and glycosphingolipids.

---

185. SUBJECT AREA: Intracellular Sorting and the Maintenance of Cellular Compartments

QUESTION: All of the following are true of the transport system EXCEPT:

A: it requires specific signals to hold back proteins.
B: it requires specialization of membranes.
C: it uses vesicles of a uniform nature.
D: mutants in the system have been characterized.
E: the transport can be studied *in vitro*.

Learning Response C: Correct. This process uses a system of specialized membranes and vesicles to transport proteins through the cell. Retained proteins must be held back from the default progression by a specific signal. The system has been studied through the use of mutants, and through *in vitro* characterization.

---

186. SUBJECT AREA: Intracellular Sorting and the Maintenance of Cellular Compartments

QUESTION: The following are characteristics of transport of secretory granules from the Golgi apparatus to the cell membrane, EXCEPT:

A: it proceeds through the budding of membrane.
B: it represents the default secretory pathway.
C: it requires more control in polarized cells.
D: the flow requires a specific signal.
E: the secretion can require a signal from the membrane.

Learning Response B: Correct. The flow of molecules to secretory granules requires a specific recognition to withdraw them from the constitutive flow to the plasma membrane. They bud from the Golgi, and in polar cells, must be directed toward the appropriate membrane. They are often stimulated to release their contents by an increase in calcium.

187. SUBJECT AREA: Intracellular Sorting and the Maintenance of Cellular Compartments

QUESTION: Viruses use cellular processes to perform all of the following functions, EXCEPT:

A: insertion of the genome through the membrane.
B: internalization of the virus.
C: sorting to the appropriate polar membrane.
D: transfer of the envelope protein to the plasma membrane.
E: translation of the capsid.

Learning Response A: Correct. Viruses will use the established processes to be internalized, to produce their soluble and membrane proteins, and have those membrane proteins transferred to the correct region of the plasma membrane. This allows the capsid to assemble around the genome in the cytoplasm, and to bud off within the envelope containing membrane.

---

188. SUBJECT AREA: Intracellular Sorting and the Maintenance of Cellular Compartments

QUESTION: Oligosaccharide chains

A: are attached in the Golgi apparatus.
B: are both pared and extended.
C: are joined to proteins through O-linkages.
D: are modified in the secretory vesicles.
E: face the cytoplasm.

Learning Response B: Correct. Oligosaccharides are attached to an asparagine in the lumenal face of the endoplasmic reticulum. They are then processed there and in the Golgi apparatus, with sugars being both added and removed.

---

189. SUBJECT AREA: Intracellular Sorting and the Maintenance of Cellular Compartments

QUESTION: The addition of GlcNAc and the removal of mannose occur in:

A: the endoplasmic reticulum.
B: the Golgi cis cisternae.
C: the Golgi medial cisternae.
D: the Golgi trans cisternae.
E: the trans Golgi network.

Learning Response C: Correct. These enzymatic reactions occur in the medial cisternae, prior to the transport of the glycosylated protein to the trans cisternae, where Gal and NANA are added. It then passes into the TGN to be sorted.

---

190. SUBJECT AREA: Intracellular Sorting and the Maintenance of Cellular Compartments

QUESTION: The production of proteoglycans involves all of the following EXCEPT:

A: addition to serines and threonines.
B: glycosyl transferase enzymes.
C: N-linked glycosylation.
D: sequential addition of sugars.
E: sulfation.

Learning Response C: Correct. The formation of proteoglycans involves O-linked glycosylation of serines and threonines. A glycosyl transferase sequentially attaches as many as ten or more sugars, which can be heavily sulfated after addition.

---

191. SUBJECT AREA: Intracellular Sorting and the Maintenance of Cellular Compartments

QUESTION: Lysosomes can receive material through all of the following, EXCEPT:

A: autophagocytic vesicles.
B: endocytosis of small molecules.
C: phagocytosis of large material.
D: secretory vesicles.
E: transport from the Golgi through endolysosomes.

Learning Response D: Correct. Endosomes containing endocytosed material are sorted, and then receive lysosomal material from the Golgi. This intermediate compartment, the endolysosome, will develop directly into a lysosome, or can fuse with phagosomes or autophagosomes.

---

192. SUBJECT AREA: Intracellular Sorting and the Maintenance of Cellular Compartments

QUESTION: All of the following are true of lysosomal hydrolases, EXCEPT:

A: they are post-translationally modified in the cis cisternae of the Golgi apparatus.
B: they are synthesized in the endoplasmic reticulum.
C: they bind a specific vesicular receptor.
D: they bud from the trans cisternae of the Golgi apparatus.
E: they require basic pH to be released into the endolysosome.

Learning Response E: Correct. These lysosomal proteins are produced in the ER, and have mannose-6-phosphate attached in the cis cisternae. This modification is recognized by specific receptors in the trans cisternae, and coated vesicles then transport the proteins to the endolysosomes. Here, the acidic pH allows dissociation, and receptor recycling.

---

193. SUBJECT AREA: The Cell Nucleus

QUESTION: All of the following are characteristics of the nucleus EXCEPT:

A: it contains two centrioles.
B: it is free of cytoskeletal filaments.
C: it is free of organelles.
D: it is the site of RNA processing.
E: it is unique to eucaryotes.

Learning Response A: Correct. The eucaryotic nucleus is thought to have developed in part to protect the DNA of the actions of microfilaments, and to confine the events of mRNA production. It is devoid of any kind of organelles, but does have a specialized nucleolus.

---

194. SUBJECT AREA: The Cell Nucleus

QUESTION: The purpose of a telomere is to:

A: bind to the nuclear membrane.
B: circularize chromosome by recognizing opposite telomere.
C: initiate replication of the chromosome.
D: pair with a sister chromatid during mitosis.
E: protect the end of the chromosome during replication.

Learning Response E: Correct. The telomere contains a repeating sequence at the end of the chromosome, which through periodic addition, protects the end of the chromosome from decrease in length, and eventual loss of genetic material through successive rounds of replication.

---

195. SUBJECT AREA: The Cell Nucleus

QUESTION: All of the following are true of eucaryotic genes EXCEPT:

A: the regions critical for function are conserved between species.
B: they are flanked by regulatory regions.
C: they contain sequence not used to code for proteins.
D: they produce functional RNA messages.
E: they vary little in size.

Learning Response E: Correct. Eucaryotic genes contain the sequence necessary for the production of a functional RNA, as well as additional regulatory regions and intervening DNA, which are not found in the mature message. The length is highly variable, depending primarily on the number and length of these insertions in the coding sequence. Usually, only the important coding and regulatory regions are conserved.

196. SUBJECT AREA: The Cell Nucleus

QUESTION: All of the following are true concerning the binding of proteins to DNA, EXCEPT:

A: it can be effected by a buffer's salt concentration.
B: it can be used to purify and study the protein.
C: it involves specific interactions with the DNA backbone.
D: it is selective for certain DNA sequences.
E: it results in a change in the migration of the DNA in a gel.

Learning Response C: Correct. The interaction of proteins with DNA involves specific interactions with the nucleotide bases and non-specific phosphate backbone contacts. The binding of the protein to DNA is dependent on salt concentration, and can be used to purify or study the protein using chromatography or gel mobility shift assays.

---

197. SUBJECT AREA: The Cell Nucleus

QUESTION: The λ cro repressor uses the following to bind DNA, EXCEPT:

A: a helix-turn-helix motif.
B: a protein homodimer.
C: a symmetric binding site.
D: a zinc finger.
E: specific base contacts.

Learning Response D: Correct. The λ cro repressor is a dimer of identical subunits, which makes specific base contacts with a symmetric (palindromic) DNA sequence. These contacts are made with a helix-turn-helix motif (rather than a zinc finger, leucine zipper, or other motif).

---

198. SUBJECT AREA: The Cell Nucleus

QUESTION: Which of the following accurately describes the flexibility of DNA?

A: DNA is always inflexibly linear.
B: DNA bending can not be changed by protein binding.
C: DNA normally adopts a bent structure.
D: protein binding to DNA reduces its flexibility.
E: some DNA sequences are inherently bent.

Learning Response E: Correct. The DNA double helix is normally slightly flexible. Certain sequences will make it more flexible, or induce a permanent bend in the DNA. Protein binding to the DNA can cause significant bending of the molecule.

---

199. SUBJECT AREA: The Cell Nucleus
QUESTION: The structure of the nucleosome core particle is NOT dependent on the presence of

A: histone H1.
B: histone H2A.
C: histone H2B.
D: histone H3.
E: histone H4.

Learning Response A: Correct. The nucleosome core is made up of histones H2A, 2B, 3 and 4. Histone H1 binds DNA between nucleosome cores.

---

200. SUBJECT AREA: The Cell Nucleus
QUESTION: Which of the following is true about the interaction between nucleosomes and DNA?

A: they are evenly distributed throughout the genome.
B: they assist in the formation of more highly compacted DNA structures.
C: they must be removed from a gene to allow it to be transcribed.
D: they require histone H1 to bind DNA.
E: they show no preference for particular sequences.

Learning Response B: Correct. The binding of nucleosomes to DNA is selective, and in some cases specific. Other regions of the DNA are not bound. With the assistance of histone H1, they can be compacted further. They need not be completely displaced, but are probably disrupted somewhat, to allow transcription.

---

201. SUBJECT AREA: The Cell Nucleus
QUESTION: All are true of mitotic chromatids EXCEPT:

A: they are covered with ribonucleoproteins.
B: they are in their most condensed state.
C: they are organized in loops of chromatin.
D: they continue to produce RNA.
E: their formation is promoted by histone H1 phosphorylation.

Learning Response D: Correct. The compaction of mitotic chromosomes into their most condensed state involves several levels of structure and can be regulated by phosphorylation of histones. It is likely that the DNA is wound so tight that RNA polymerase can not gain access to the DNA.

---

202. SUBJECT AREA: The Cell Nucleus
QUESTION: All the following are true of the banding pattern seen in chromosomes EXCEPT:

A: it can only be observed in mammalian chromosomes.
B: it is conserved across related species.
C: it is produced by the binding of selective dyes.
D: it produces a ready means for identifying chromosomes.
E: it results from differences in the nucleotide content.

Learning Response A: Correct. A banding pattern is observed in chromosomes when dyes selected for certain nucleotides are used as a stain. They can be observed in many species including humans and flies and their pattern enables the identification of analogous regions in the chromosomes of related species.

---

203. Subject Area: The Cell Nucleus

Question: The following are true of polytene chromosomes, EXCEPT:

A: they are observed in *Drosophila* larval salivary glands.
B: they can be observed with the light microscope.
C: they can unfold and refold as a unit.
D: they cannot transcribe RNA.
E: they contain over 1000 identical strands.

Learning Response D: Correct. Polytene chromosomes, as observed in *Drosophila* salivary glands, are the result of numerous rounds of DNA replication without cell division. They can be observed with a light microscope as they are seen to unfold as puffs in order to transcribe RNA, and then return to their original banded structure.

---

204. Subject Area: The Cell Nucleus

Question: All the following are true of active chromatin EXCEPT:

A: it binds histone H1 more tightly.
B: it can be transcribed.
C: it is bound to highly acetylated nucleosomes.
D: it is less condensed.
E: it is more easily degraded by DNase.

Learning Response A: Correct. Active chromatin has a more open configuration and hence is more able to be transcribed but is also more susceptible to DNase. This configuration is due to biochemical differences in the proteins responsible for chromatin structure including highly acetylated nucleosomes and a histone H1 which binds less tightly.

---

205. Subject Area: The Cell Nucleus

Question: Eucaryotic chromosome replication begins:

A: at clusters of sites along the length of the chromosome.
B: at random sites along the chromosome.
C: at the centromere and progresses toward the ends.
D: at the end of the chromosome and moves toward the center.
E: at the single origin of replication in each arm of the chromosome.

Learning Response A: Correct. Replication begins at specific replication origins located in clusters called replication units, spaced unevenly along the chromosome.

---

206. SUBJECT AREA: The Cell Nucleus
QUESTION: All the following are true of telomeres EXCEPT:

A: they are located at the end of chromosomes.
B: they are replicated without a primer.
C: they consist of repeating sequence units.
D: they fold to form a special structure.
E: they have the same sequence in all species.

Learning Response E: Correct. Telomeres fold to produce a capping structure at the end of chromosomes. They contain a short repeating sequence, differing among species, that is extended in a primer independent manner.

---

207. SUBJECT AREA: The Cell Nucleus
QUESTION: Which of the following are known to be late replicating regions?

I. A-T-rich bands.
II. active chromatin.
III. highly condensed chromatin.

A: I.
B: II.
C: III.
D: I and III.

E: II and III.

Learning Response D: Correct. Replication appears to begin near regions of active chromatin, while regions of highly condensed chromatin and those corresponding to the A-T-rich bands seen at metaphase replicate late in S phase.

---

208 SUBJECT AREA: The Cell Nucleus
QUESTION: All of the following are true of ribosomal RNA EXCEPT:

A: its assembly into a ribosome occurs in the nucleolus.
B: it is cleaved in the nucleus.
C: its gene can be seen in chromatin preparations.
D: its gene exists as a single copy in the haploid genome.
E: its maturation requires it to complex with different polypeptide chains.

Learning Response D: Correct. The ribosomal RNA is produced as a single transcript from many gene copies. It is processed into three smaller RNAs which are assembled with proteins in the nucleolus to produce ribosomes.

---

209. SUBJECT AREA: The Cell Nucleus
QUESTION: Which of the following is a characteristic of the nucleolus?

A: it forms randomly after mitosis.
B: it is bound by a membrane.
C: it is of constant size.
D: it remains constant over the cell cycle.
E: several regions can be distinguished in it.

Learning Response E: Correct. The nucleolus is the center for ribosome production, in which three regions can be distinguished. Its activity can be reflected by its size. It disappears during mitosis, but reforms on the same specific chromosomes.

210. SUBJECT AREA: The Cell Nucleus
QUESTION: All the following are true of chromosomal arrangement within nuclei EXCEPT:

A: chromosomes are oriented during mitosis.
B: chromosomal arrangement has most often been studied using polytene chromosomes.
C: chromosomal arrangement is random in human cells.
D: in some cases chromosomes are specifically oriented.
E: the orientation of chromosomes can not be easily distinguished during interphase.

Learning Response C: Correct. Chromosomal arrangement in the nucleus is difficult to study except for certain exceptions. A special arrangement has been observed in mitotic cells, in polytene chromosomes and through the assembly of the nucleolus in human cells.

---

211. SUBJECT AREA: The Cell Nucleus
QUESTION: Which of the following is true of hnRNA transcripts?

A: they are immediately bound with proteins.
B: they are immediately exported from the nucleus.
C: they are rapidly degraded.
D: they have had their introns removed.
E: they form snRNPs.

Learning Response A: Correct. Eucaryotic RNA transcripts are immediately bound into hnRNP particles. They are then spliced through the action of snRNPs, before they are exported from the nucleus.

---

212. SUBJECT AREA: The Cell Nucleus
QUESTION: The spliceosome which recognizes the 5' splice site during RNA processing is:

A: Ul.

B: U2.
C: U5.
D: U4/6.
E: none of the above.

Learning Response A: Correct. The U1 spliceosome recognizes the 5' (donor) site, interacting with the U2 spliceosome binding the branch site, and allowing the formation of the lariat.

---

213. SUBJECT AREA: The Cell Nucleus
QUESTION: A mutation of G to A in the consensus GT splice donor site (GT — AG) located at the first (5') intron-exon junction of the insulin gene would likely result in

A: an altered rate of transcription of new precursor mRNA molecules from the mutant insulin gene.
B: an increased stability of insulin mRNA.
C: faulty polyadenylation of the insulin pre-mRNA.
D: possible use of a "cryptic" alternative splice donor site within the intron sequence.
E: unaltered cytoplasmic levels of normal insulin mRNA.

Learning Response D: Correct. While mutation of a splice site has no effect on prior (transcription) or unconnected (polyadenylation) steps in processing, it results in abnormal mRNA molecules spliced at cryptic sites, as in the thalassemias, or unstable messages.

---

214. SUBJECT AREA: The Cell Nucleus
QUESTION: What explains the apparent shortening of eucaryotic RNA transcripts following their synthesis?

A: it is an artifact of the experiment.
B: they are cleaved into several messages.
C: they are degraded from the 3' end.
D: they are degraded from the 5' end.
E: they have regions removed from the middle.

Learning Response E: Correct. The following transcription regions of intervening sequence are removed from within the RNA in a process referred to as splicing.

---

215. SUBJECT AREA: The Cell Nucleus

QUESTION: Which of the following statements about the eucaryotic polymerases is correct?

A: each transcribes the same genes.
B: polymerase I produces mRNA.
C: polymerase II produces most of the snRNA.
D: polymerase III is primarily involved in rRNA synthesis.
E: they use all the same transcription factors.

Learning Response C: Correct. The three polymerases found in eucaryotes have different functions, and are activated by unique combinations of transcription factors. Polymerase I and polymerase III produce the small rRNA and small stable RNAs, respectively. Polymerase II produces the RNA which is transcribed into protein and also the RNAs that form snRNPs.

---

216. SUBJECT AREA: The Cell Nucleus

QUESTION: All the following are true of bacterial RNA polymerase action EXCEPT:

A: a single polymerase is used.
B: it binds a sigma subunit before interacting with the DNA.
C: it exchanges subunits after initiating transcription.
D: it exists as a five subunit complex.
E: the rate of initiation is affected by regulatory proteins.

Learning Response D: Correct. Polymerase activity in procaryotes is carried out by a single enzyme made of six subunits. Before initiating it must bind a sigma factor which enables it to initiate. This is exchanged for elongation factors after elongation has occurred. The binding of polymerase to the promoter is regulated by proteins bound to adjacent sites.

217. SUBJECT AREA: Control of Gene Expression

QUESTION: Which of the following statements accurately portrays gene regulatory proteins?

A: a cell's history will play a role in determining the gene's regulatory proteins present.
B: a single gene regulatory protein will have the same effect on every gene that it regulates.
C: all cells express the same gene regulatory proteins.
D: each cell type has a characteristic gene regulatory protein.
E: each gene is usually controlled by a single regulatory protein.

Learning Response A: Correct. The expression of genes in cells depends on the complex interaction of gene regulatory proteins. These show different expression in different cell types, and at different times in development. It is the interplay of the various regulatory proteins that determine which genes are to be expressed, including additional regulatory proteins. In this way the expression of genes depends on the cell's history of expressing regulatory proteins.

218. SUBJECT AREA: Control of Gene Expression

QUESTION: Differences in the types of proteins expressed in different cells are:

A: due solely to differences in gene transcription.
B: easily studied by a direct comparison between total protein in cells.
C: more pronounced than the similarities in expression.
D: responsible for the differences seen between cells.
E: usually due to differences in their DNA.

Learning Response D: Correct. Differences in gene transcription, RNA processing, transport, translation and stability, result in differences in proteins in spite of cells having the same DNA. These often subtle differences in protein expression can result in cells with very different characteristics.

219. SUBJECT AREA: Control of Gene Expression

QUESTION: Transcription initiation can be controlled by feedback through the following mechanisms, EXCEPT:

A: binding of a product to a stimulatory factor to prevent its interaction with DNA.

B: binding of an RNA product to the gene to prevent transcription.

C: facilitated binding of a repressor by the product of the pathway in which the gene product is involved.

D: removal of a repressor in the presence of the substrate for a pathway in which the gene product is involved.

E: removal of a substrate which must bind a stimulatory factor to allow gene activity.

Learning Response B: Correct. While there are many direct and indirect mechanisms for the feedback control of transcription, including the various interactions of substrates and products with transcription factors and the interaction of RNA products, the interaction of RNA products with DNA to inhibit transcription has not been observed.

220. SUBJECT AREA: Control of Gene Expression

QUESTION: Which of the following statements about transcriptional regulators is correct?

A: a DNA binding protein can only stimulate transcription by interacting with RNA polymerase.

B: a repressor can only work by blocking RNA polymerase binding.

C: an activator must work by contacting DNA directly.

D: for a protein to effect RNA polymerase, its binding site must be immediately adjacent to the start site.

E: for a protein binding a DNA sequence to effect RNA polymerase, the site must be linked to the start site.

Learning Response E: Correct. To effect the initiation of transcription by

RNA polymerase, a protein must either bind a site that is linked to the start site, or interact with a protein that is bound to the DNA. This binding can be at a distance, due to the flexibility of DNA. The protein then affects RNA polymerase by direct interaction, or by interacting with another protein which in turn interacts with polymerase. A repressor can mediate its effects by blocking any of these interactions.

---

221. SUBJECT AREA: Control of Gene Expression
QUESTION: Domain swapping experiments using transcription factors reveals:

A: that chimeric proteins cannot function as transcription factors.
B: that the DNA binding domain of a factor is relatively non-specific.
C: that the domains of a transcription factor are intrinsically linked to each other.
D: that the domains of transcription factors are somewhat modular.
E: none of the above.

Learning Response D: Correct. The domains of a transcription factor are somewhat distinct, and can be separated and recombined with other factors without losing their effect.

---

222. SUBJECT AREA: Control of Gene Expression
QUESTION: The activity of transcription factors can be controlled by:

A: adjusting their relative concentration in the cell.
B: covalently modifying the protein.
C: interactions with other proteins.
D: the binding of a ligand.
E: all of the above.

Learning Response E: Correct. Interactions with other proteins, covalent modification (such as phosphorylation), and ligand binding can affect either the ability of a protein to bind, or to have its effect on RNA polymerase once bound. By altering the synthesis or degradation of a protein, or by competing for a binding site with another protein, the amount of protein bound to the DNA can also be affected. All of these schemes are used to regulate activity.

---

223. SUBJECT AREA: Control of Gene Expression
QUESTION: The classical definition of an enhancer included all of the following criteria, EXCEPT:

A: the ability to act at a distance as long as they are linked to the start site.
B: the ability to directly interact with RNA polymerase.
C: the ability to function in either orientation.
D: the ability to work either upstream or downstream of the start site.
E: all of the above are required to identify the element as an enhancer.

Learning Response B: Correct. While these criteria are no longer adhered to as strictly, the classical enhancer had to work independent of the distance (within limits), of their orientation, of their location relative to the start site, or even of the promoter itself. However, it need not interact directly with RNA polymerase, as this characterization predates an understanding of mechanism.

---

224. SUBJECT AREA: Control of Gene Expression
QUESTION: Gene conversion tends to:

A: be unaffected by the chromosomal location of the genes.
B: eliminate extra copies of genes.
C: homogenize the sequences of related genes.
D: increase the rate of sequence divergence.
E: prevent gene duplication.

Learning Response C: Correct. The effect of gene conversion is to maintain similar or even identical sequences in closely related genes. This results in a lower rate of divergence between duplicated genes than would be expected.

These effects seem to be minimized when the genes are moved to different chromosomes.

---

225. SUBJECT AREA: Control of Gene Expression

QUESTION: Ancestral genes constructed from numerous shuffled exons would predict all of the following, EXCEPT:

A: decreased recombination events than in intron-less genes.
B: exons containing single protein domains.
C: shared exon boundaries among distantly related proteins.
D: similar protein domains found in different proteins.
E: all of the above would be predicted.

Learning Response A: Correct. The separate domain structure of these proteins would allow them to be placed together more easily (by recombination events within the intervening sequences) than a protein without such domain structure, where a random recombinatorial event would be likely to disrupt a domain. This would allow a limitless number of combinations to arise, resulting in proteins with a domain structure, and sharing common domains. Only after a highly efficient protein had arisen might the elimination of the introns be favored, resulting in only a subset of the introns being preserved among any two distantly related proteins.

---

226. SUBJECT AREA: Control of Gene Expression

QUESTION: Which of the following statements about transposons is correct?

A: they all have inverted repeats at the ends.
B: they all move through an RNA intermediate, and encode reverse transcriptase.
C: they excise themselves without effecting the gene into which they have inserted.
D: they can only exist as a single copy in any one cell.
E: they can transport a region of genomic DNA with them if inappropriately excised.

Learning Response E: Correct. There are several different classes of transposons, with differing characteristics, including nature of their ends and replication intermediate. Many copies of each can exist in cells at a time, and they can cause mutations by altering their insertion site upon excision, or by moving a region of a gene with them to a new location.

---

227. SUBJECT AREA: Control of Gene Expression

QUESTION: Satellite DNA sequences share all of the following characteristics, EXCEPT:

A: are highly conserved among individuals.
B: can be separated from other DNA because of their unusual nucleotide ratio.
C: can form an entire chromosome arm.
D: comprise the heterochromatic regions of the chromosome.
E: contain short sequences repeated many times.

Learning Response A: Correct. Satellite DNA appears to be a form of selfish DNA, with many copies of a highly repetitive sequence which represent as much as 10% of the total DNA. They demonstrate great changes in their arrangement among individuals, and thus can be used for typing individuals.

---

228. SUBJECT AREA: Control of Gene Expression

QUESTION: All of the following are mechanisms of posttranscriptional regulation, EXCEPT:

A: alternative splicing of RNA transcripts.
B: base pairing between RNA molecules.
C: blocking translation by binding of a protein to the RNA.
D: regulation of RNA half life by protein binding.
E: all of the above are examples of posttranscriptional regulation.

Learning Response E: Correct. Posttranslational regulation includes all of the above mechanisms, as well as frameshifts in translation, premature termination, alternative polyadenylation, and other mechanisms.

229. SUBJECT AREA: Control of Gene Expression

QUESTION: The switch which regulates premature termination of a transcript in bacteria involves all of the following, EXCEPT:

A: an alternative structure adjacent to a termination signal.
B: base pairing between two different RNA transcript.
C: formation of RNA stem-loop structures.
D: genes which play a role in amino acid synthesis.
E: protein binding to the RNA molecule being transcribed.

Learning Response B: Correct. This process involves the formation of alternative stem-loop structures in a single RNA transcript. One of these structures results in a stem immediately next to a poly-U sequence, which structure serves as a bacterial termination signal. This is prevented by the binding of a protein, often a stalled ribosome, over one of the sequences involved in the non-permissive structural arrangement.

230. SUBJECT AREA: Control of Gene Expression

QUESTION: Which of the following is NOT true concerning alternative splicing of RNA?

A: it allows for more than one protein to be made from a single mRNA.
B: it can result in an RNA transcript which does not encode an active protein.
C: it can result in protein isoforms with alternative domains.
D: it can result in proteins with vastly differing sequences.
E: it need not be a regulated event.

Learning Response A: Correct. Alternative splicing can have a broad range of effects on a protein, from a product with no activity, to one with only an added domain. By the introduction of a difference in reading frame, or of an alternative poly-adenylation site, proteins of very different characteristics can be produced from the same RNA transcript. However, once spliced into an mRNA molecule, only a single protein product would be expected.

---

231. SUBJECT AREA: Control of Gene Expression

QUESTION: Binding of a regulatory protein to the 5'-untranslated region of a mRNA:

A: increases the rate of translation.
B: is blocked by a translational enhancer sequence.
C: is limited to bacterial systems.
D: mediates its effect primarily on message half-life.
E: usually occurs at a stem-loop structure.

Learning Response E: Correct. Protein binding at a stem-loop at the 5' end of a mRNA results in a repression of the translation of the message. The mechanism is found in humans, as well as many other species.

---

232. SUBJECT AREA: Control of Gene Expression

QUESTION: A change in the reading frame of an RNA has been characterized in which of the following?

I. a translational frameshift in viral mRNA.
II. alternative splicing of RNA transcript.
III. insertion of nucleotides into mRNA.
IV. transposon excision from an RNA transcript.

A: I and II.
B: I and III.
C: I and IV.
D: I, II and III.
E: I, II and IV.

Learning Response D: Correct. Alternative splicing and mRNA editing have been observed, the latter with bases both inserted and deleted from specific transcripts. Transposon excision will only occur from the DNA, prior to transcription, where it could potentially produce a frameshift.

233. SUBJECT AREA: Control of Gene Expression
QUESTION: DNA methylation in mammalian cells:

A: can be preserved by passing through bacterial plasmids.
B: converts uracil to thymidine.
C: is inherited through several rounds of replication.
D: is used to distinguish between strands.
E: tends to be more frequent in active genes.

Learning Response C: Correct. In mammalian cells, DNA is methylated on the C residues of CG pairs. This is propagated through rounds of replication in mammalian cells, but not in bacteria. It seems to be less frequent in actively transcribed regions of the genome.

234. SUBJECT AREA: Control of Gene Expression
QUESTION: Which of the following statements about yeast mating types is correct?

I. a switch in mating type occurs with each division.
II. all yeast have both mating type genes.
III. as part of the switching process, a double stranded break is made in the DNA.
IV. the mating type is inherited only by the daughter cell.

A: I and III.
B: I and IV.
C: I, III and IV.
D: II and III.
E: II, III and IV.

Learning Response D: Correct. Yeast have inactive genes for each of the mating types which are used in an alternating fashion to replace the active gene when a specific double stranded break has been made during a switch. The mating type is switched at the time of a mother cell division, and both mother and daughter inherit the type. However, the daughter does not switch until it produces its own mother cell through a round of budding.

235. SUBJECT AREA: Control of Gene Expression
QUESTION: Which of the following accurately describes an aspect of X inactivation?

A: cells with different X chromosomes inactivated are randomly distributed.
B: fusion chromosomes do not inactivate.
C: inactivation proceeds by a cooperative nucleation process.
D: no genes on the inactive X chromosome are inactivated.
E: X inactivation occurs randomly after every division.

Learning Response C: Correct. X inactivation occurs between the third and sixth days of development, and is faithfully inherited during subsequent cell divisions, resulting in a clustering of cells with the same X inactivated. It begins at a nucleation site and proceeds through the adjacent chromosome, even in fusions, and leaves only a few genes still active.

---

236. SUBJECT AREA: Control of Gene Expression
QUESTION: Which of the following statements regarding phase variation in bacteria is NOT true?

A: it involves different types of the same protein.
B: it involves the movement of genes away from the promoter.
C: it reverses the orientation of the promoter.
D: it uses site-specific recombination.
E: all of the above are true.

Learning Response B: Correct. The process of phase conversion involves the inversion of a region containing the promoter for both a flagellin gene and a repressor of the expression of a different flagellin. The inversion of this region thus results in the loss of one type and the expression of the other.

---

237. SUBJECT AREA: Control of Gene Expression

QUESTION: All of the following describe the λ bacteriophage switch, EXCEPT:

A: it involves reversible changes in the DNA sequence.
B: it relies on only two gene products.
C: the regulation is different depending on the condition of the host cell.
D: the virus uses it to regulate its lysogenic/lytic states.
E: all of the above are true.

Learning Response A: Correct. The λ switch involves two proteins that each repress the transcription of the other (as well as of other genes). The balance between these two controls whether the virus will remain lysogenic, or become lytic, depending on the condition of the host cell.

---

238. SUBJECT AREA: Control of Gene Expression
QUESTION: Genetic switches in mammals

A: are entirely different even among related species.
B: do not resemble those seen in bacteriophage.
C: involve only transcriptional repression.
D: usually involve larger networks of regulatory inputs.
E: none of the above are true.

Learning Response D: Correct. While some basic similarities can be seen, the regulation of eucaryotic genes is usually more complex than that of bacteriophage. It usually involves multiple positive and negative elements, some of which can be identified by comparing the DNA of related species for conservation.

---

239. SUBJECT AREA: Control of Gene Expression
QUESTION: All of the following are true of the rapid degradation of RNA in eucaryotic cells, EXCEPT:

A: it can be inhibited by protein binding to the 3' end of the RNA.

B: it can be regulated by specific structures in the RNA.
C: it can be transferred by exchanging regions of the RNA.
D: it is blocked by protein synthesis inhibitors.
E: it is inhibited by A/U rich sequences in the RNA.

Learning Response E: Correct. The presence of stretches of A/U rich sequence in the mRNA, unlike the binding of proteins there, tends to increase the instability of the RNA.

---

240. SUBJECT AREA: Control of Gene Expression
QUESTION: The presence of CG sequences

A: form islands around non-transcribed regions.
B: identify regions without "housekeeping genes".
C: if methylated, is maintained over evolutionary time.
D: is reduced in vertebrates, compared to invertebrates.
E: is decreased through the deamination of the methylated C to U.

Learning Response D: Correct. Since methylated sequences are more common in regions not being transcribed, these regions are liable to lose these sequences. The deamination of methylated C results in its replacement by T. Over time, this selective loss of methylated GC sequences results in the formation of islands of (non-methylated) CG sequences surrounding actively transcribed genes, and specifically those "housekeeping genes" active in all tissues.

---

241. SUBJECT AREA: The Cytoskeleton
QUESTION: Which of the following statements about myofibrils is true?

A: each comprises a single cell.
B: they are 1-2m in length and often the same diameter as a muscle fiber.
C: they are composed of thick and thin filaments.

D: they expand in the presence of ATP and calcium.
E: they represent a relative minority of the mass of a muscle fiber.

Learning Response C: Correct. A muscle fiber is a large multinucleated single cell, a majority of which is occupied by myofibrils. These may be as long as the muscle fiber, and contract in the presence of ATP and calcium.

---

242. SUBJECT AREA: The Cytoskeleton
QUESTION: During skeletal muscle contraction, which of the following will become shortened?

I. the dark band.
II. the light band.
III. the sarcomere.
IV. the Z disc.

A: I and II.
B: III and IV.
C: I, II and III.
D: II and IV.
E: I and IV.

Learning Response C: Correct. During muscle contraction, the dark band (A band) representing the myosin thick filament, will remain the same size. By sliding along the actin filaments to shorten both the light band (I band) and the entire sarcomere.

---

243. SUBJECT AREA: The Cytoskeleton
QUESTION: All of the following are properties of actin thin filaments EXCEPT:

A: the actin is in the form of a helix.
B: the filaments are polar in nature.
C: they can form *in vitro*.
D: they result from the polymerization of F actin to form G actin.

E: all of the above are true.

Learning Response D: Correct. Globular (G) actin will polymerize *in vitro* to form filamentous (F) actin, which filaments are both helical and polar.

---

244. SUBJECT AREA: The Cytoskeleton

QUESTION: Which of the following statements accurately depicts a property of myosin?

A: it can be cleaved into light and heavy chains by papain.
B: it exists as a coiled-coil dimer of light and heavy chains.
C: it requires actin for ATPase activity.
D: it requires energy to assemble into filaments.
E: none of the above are true.

Learning Response E: Correct. Myosin is a multi-subunit protein composed of two heavy chains and four light chains, that will assemble spontaneously in vitro, and have a low but detectable ATPase activity in the absence of actin. The heavy chains are linked by a coiled-coil interaction, and can be cleaved by papain into heads and rods.

---

245. SUBJECT AREA: The Cytoskeleton

QUESTION: The process of contraction involves:

A: a drop in the level of calcium.
B: the binding of calcium to tropomyosin to release its inhibition of myosin activity.
C: the exchange of ATP for ADP, which allows the myosin to bind the actin filament.
D: the motion of an actin filament toward the Z disk.
E: the release of actin by troponin I.

Learning Response E: Correct. An increase in calcium causes troponin C to remove troponin I from its interaction with actin. This allows the myosin to bind to the actin, and pull itself along the actin filament, and toward the Z disc. The exchange of ATP for ADP then allows the myosin to release from the actin.

---

246. SUBJECT AREA: The Cytoskeleton

QUESTION: What aspect of heart muscle differs from skeletal muscle?

A: it contains intercalating discs.
B: it has a different myosin/actin structure.
C: it has multinucleate cells.
D: it has no Z disc.
E: its contraction is independent of calcium.

Learning Response A: Correct. Heart muscle is made up of mononucleate cells, connected through gap junctions, with the intercalating disc serving the same function, the linking of sarcomeres between cells, as the Z disc serves within cells.

---

247. SUBJECT AREA: The Cytoskeleton

QUESTION: Myosin/actin mediated contraction in non-muscle cells:

A: does not occur.
B: does not require ATP.
C: is not affected by calcium.
D: is promoted by the phosphorylation of myosin heavy chain.
E: only occurs transiently for specific functions.

Learning Response E: Correct. In non-muscle cells, a more indirect mechanism regulates the transient formation of myosin/actin contractile assemblages. It leads to the phosphorylation of the myosin light chain, which allows the myosin to form short filaments, which hydrolyze ATP at a rate similar to smooth muscle.

---

248. SUBJECT AREA: The Cytoskeleton

QUESTION: All of the following mediate their effects on the cytoskeleton through their action on actin EXCEPT:

A: gelsolin
B: filamin
C: fimbrin
D: myosin
E: all of the above mediate their effects through actin.

Learning Response E: Correct. Filamin, by cross-linking actin filaments, allows the formation of a gel which can be disrupted by gelsolin, in the presence of calcium. Myosin pulls one actin filament against another, while fimbrin is an actin-bundling protein.

---

249. SUBJECT AREA: The Cytoskeleton
QUESTION: In red blood cells, spectrin serves to:

A: block the interaction of actin and myosin.
B: cap the end of an actin filament.
C: help to support the plasma membrane.
D: link ankyrin to the plasma membrane.
E: stabilize microvilli.

Learning Response C: Correct. Spectrin tetramers, which are bound at their ends to short actin filaments, provide support to the plasma membrane through their ankyrin link to the 'band 3' trans-membrane protein.

---

250. SUBJECT AREA: The Cytoskeleton
QUESTION: All of the following are thought to be involved in the linkage of an actin filament to the plasma membrane at a site of attachment to the extracellular matrix EXCEPT:

A: α-actinin.
B: fibronectin.
C: fibronectin receptor.
D: talin.
E: vinculin.

Learning Response B: Correct. Actin is thought to be linked through a capping protein and α-actinin, to a complex on the membrane composed of vinculin, talin, and the trans-membrane fibronectin receptor. This, in turn binds fibronectin in the extracellular matrix.

---

251. SUBJECT AREA: The Cytoskeleton
QUESTION: Actin polymerization is more efficient:

A: after an initial lag phase.
B: at low concentration.
C: at the minus end.
D: when actin is ADP bound.
E: when it is bound to myosin heads.

Learning Response A: Correct. After a lag due to the slow initial nucleation, actin follows a pattern of polymerization which is more efficient at the plus end, with higher concentrations, and when ATP bound.

---

252. SUBJECT AREA: The Cytoskeleton
QUESTION: Cilia and flagella are similar to each other in that they:

A: are of similar lengths.
B: both resemble bacterial flagella.
C: have similar structures.
D: make similar movements.
E: occur on cells in similar numbers.

Learning Response C: Correct. While flagella are longer and move in a sinusoidal manner, cilia usually occur in larger numbers and move in a whip-like manner. Both have a similar structure of microtubules, which is entirely different from the structure of procaryotic flagella.

---

253. SUBJECT AREA: The Cytoskeleton
QUESTION: Which of the following are aspects of microtubule structure:

I. they are arranged in cilia as a ring of nine triplets.

II. they are composed of two different tubulin subunits.
III. they bind to microtubule accessory proteins.
IV. they occur as a cylinder with a circumference of 13 subunits.

A: I, II and III.
B: I, II and IV.
C: I, III and IV.
D: II, III and IV.
E: I, II, III, and IV.

Learning Response D: Correct. Microtubules, composed of α and β subunits arranged in a 13 subunit cylindrical structure. They occur in cilia as an array of nine doublets around two single central tubules.

---

254. SUBJECT AREA: The Cytoskeleton
QUESTION: Which of the following provides for the bending of a ciliary axoneme?

A: dynein sliding one microtubule along the next.
B: nexin pulling adjacent doublets together.
C: radial spokes pushing against central microtubules.
D: sheath projections rotating central microtubules around the center.
E: tubulin doublets sliding along each other.

Learning Response A: Correct. It is the dynein which drag the B tubule of the adjacent doublet along the A tubule to which they are bound permanently.

---

255. SUBJECT AREA: The Cytoskeleton
QUESTION: Which of the following is a property of centrioles?

A: more than a pair occur in a cell only immediately before cell division.
B: they differ in structure from basal bodies.
C: they have the same 9+2 structure as cilia.
D: they serve as focal point for the centrosome.

E: they show an increase in the number of triplets as they mature.

Learning Response D: Correct. Centrioles have a structure of nine tubules, that form into triplets as they mature. During division, they serve as focal points for centrosomes. In non-dividing cells, they can serve as basal bodies for thousands of cilia.

---

256. SUBJECT AREA: The Cytoskeleton
QUESTION: The following will effect dynamic instability of microtubules in a cell, EXCEPT:

I. catastrophic depolymerization of the plus end.
II. rapid growth at the minus end.
III. stabilization by a capping structure.
IV. the ADP/ATP bound state of the tubulin.

A: I and II.
B: 1 and III.
C: I and IV.
D: II and III.
E: II and IV.

Learning Response E: Correct. Microtubules in vivo tend to grow rapidly at their plus end, until hydrolysis of bound GTP renders them unstable, and they undergo a rapid depolymerization, that is thought to be inhibited if the plus end is protected by some capping structure.

---

257. SUBJECT AREA: Cell Signaling
QUESTION: Paracrine action involves:

A: action at a synapse.
B: hormones that act at a distance.
C: local chemical mediators.
D: secretion into the bloodstream.
E: secretion of a neurotransmitter.

Learning Response C: Correct. Paracrine action involves the action of compounds which are rapidly taken up or destroyed, and thus only act on neighboring cells. This is distinct from synaptic action, which involves a special contact between the cells sending and receiving the signal.

---

258. SUBJECT AREA: Cell Signaling

QUESTION: Compared to synaptic signaling, endocrine secretion would be expected to involve:

A: a closer interaction between the cells.
B: a higher concentration at the site of action.
C: a shorter duration of response.
D: action on a smaller number of cells.
E: action at a specific receptor.

Learning Response E: Correct. Due to the formation of synapses, signaling at synapse will involve a higher concentration at the site of action, will be more rapid to initiate, but of shorter duration, and will act on a single cell. Endocrine action involves secretion into the bloodstream, which will yield lower concentrations, slower but longer action, and will effect every cell with the appropriate specific receptor.

---

259. SUBJECT AREA: Cell Signaling

QUESTION: All of the following are examples of hormones EXCEPT:

A: amino acid derivatives, including epinephrine.
B: peptides, including vasopresin.
C: proteins, including somatotropin.
D: steroids, including testosterone.
E: all of the above are examples of hormones.

Learning Response E: Correct. Hormones belong to many classes ranging in size from small amino acid derivatives and cholesterol-based steroids through small peptides to large heteromeric, glycosylated proteins.

---

260. SUBJECT AREA: Cell Signaling

QUESTION: Which of the following statements about hormones is true?

A: all hormones act at plasma membrane receptors.
B: hydrophilic hormones must bind a carrier protein in the plasma membrane.
C: hydrophobic hormones tend to be rapidly removed from the circulation.
D: lipid-soluble hormones bind receptors within the cell.
E: protein hormones act by passing through the cell membrane.

Learning Response D: Correct. Hydrophobic (lipid-soluble) hormones such as steroids and thyroid hormone pass through the membranes of the cell and bind to receptors which reside in the cytoplasm or nucleus. They tend to have significantly longer half lives than do proteins or peptides.

---

261. SUBJECT AREA: Cell Signaling
QUESTION: Which of the following is an accurate statement about steroid hormone receptors?

A: binding of hormone allows association with heat shock proteins.
B: each activates a unique gene.
C: they act through a second messenger system.
D: they all bind to the same receptor.
E: they bind directly to DNA when bound to steroid.

Learning Response E: Correct. Each steroid hormone regulates a distinct set of genes. This is done through the binding to a unique receptor within the cell, which is some cases causes the dissociation of this receptor from associated heat shock proteins. The hormone-bound receptor is then freed to bind directly to DNA.

---

262. SUBJECT AREA: Cell Signaling
QUESTION: Features of prostaglandins as chemical mediators include all of the following EXCEPT:

A: their action is short lived, unlike most hydrophobic hormones.
B: they act through the binding of intercellular receptors.
C: they are produced in every tissue.
D: they are synthesized from a fatty acid precursor.
E: they can act on the cell which secretes them.

Learning Response B: Correct. Prostaglandins are synthesized in all tissues from arachidonic acid, and secreted continuously. They can act on the cell which secretes them (autocrine) as well as neighboring cells (paracrine), where they bind cell-surface receptors and initiate second messenger cascades. Their actions are relatively short lived.

---

263. SUBJECT AREA: Cell Signaling
QUESTION: The binding of a hormone to a cell-surface receptor might be expected to produce which of the following effects:

I. alteration in a receptor mediated catalytic activity.
II. production of a second messenger.
III. opening of a membrane ion channel.
IV. translocation of the receptor to the nucleus.

A: I, II, and III.
B: I, II, and IV.
C: I, III, and IV.
D: II, III, and IV.
E: I, II, III, and IV.

Learning Response A: Correct. While translocation of a receptor to the nucleus is a characteristic of certain intercellular hormone receptors, the actions of cell surface receptors appears to be limited to opening ion channels, initiating second messenger cascades, and altering receptor catalytic activity.

---

264. SUBJECT AREA: Cell Signaling
QUESTION: Which of the following characteristics is true of the G

protein which stimulates adenylate cyclase?

A: dissociation of the inactive α subunit allows the remaining subunits to activate the enzyme.
B: it is activated by binding the hormone with its extracellular domain.
C: it is inhibited by cholera toxin.
D: it must compete with another G protein, which inhibits the enzyme.
E: it only activates when bound to GDP.

Learning Response D: Correct. The G protein is a heterotrimer that is induced to bind GTP by its interaction with the hormone-bound receptor. This induces the dissociation of the α subunit, which then activates adenylate cyclase until it hydrolyzes its nucleotide to GDP. This hydrolysis is prevented by cholera toxin mediated ADP ribosylation, constitutively activating the G protein, and in its turn, adenylate cyclase. A G protein activated by a different class of receptors has an inhibitory effect on the enzyme.

---

265. SUBJECT AREA: Cell Signaling
QUESTION: Which of the following represent components necessary to study the activation of adenylate cyclase by pertussis toxin?

A: G_i, adenylate cyclase, pertussis toxin, ADP, ATP, GTP.
B: G_i, adenylate cyclase, pertussis toxin, ATP, FAD.
C: G_i, adenylate cyclase, pertussis toxin, ATP, GTP, NAD.
D: G_s, adenylate cyclase, pertussis toxin, ADP, GDP, FAD.
E: G_s, adenylate cyclase, pertussis toxin, ATP, GTP.

Learning Response C: Correct. In order to study the activity of pertussis toxin on adenylate cyclase, one would need, in addition to the toxin and the effector enzyme, the appropriate G protein (G_i), and the substrates for the various stages of the reaction. These are ATP, the substrate for the enzyme, NAD, used by the toxin to ADP ribosylate the G protein, and the GTP required for G protein activity.

266. SUBJECT AREA: Cell Signaling
QUESTION: The cellular homologues of the *ras* oncogene:

I. are heterotrimeric G proteins.
II. are probably identical with G_p.
III. can be activated by a single mutation.
IV. play a role in transduction of growth factor signals.

A: I and II.
B: I and III.
C: I and IV.
D: II and III.
E: III and IV.

Learning Response E: Correct. The product of the *ras* oncogene represents a constitutively activated version of a normal cellular protein, which binds to GTP, and translocates to the cell membrane as part of the signaling pathway of growth factor receptors, it is distinct from the heterotrimeric G proteins, including G_p.

267. SUBJECT AREA: Cell Signaling
QUESTION: Growth factor receptors:

A: act solely as monomers.
B: primarily activate the heterotrimeric G proteins.
C: are serine kinases.
D: have seven trans-membrane domains.
E: usually phosphorylate themselves.

Learning Response E: Correct. Growth factor receptors exist as proteins with single transmembrane domains and minimal tyrosine kinase activity. When they bind hormone, dimers of the receptor auto-phosphorylate their tyrosine kinase domains, and this activates the signal transduction cascade of the receptors, which does not involve the heterotrimeric G proteins.

268. SUBJECT AREA: Cell Signaling

QUESTION: Seven transmembrane domain receptors and related proteins stimulate G proteins which directly activate all of the following second messenger systems EXCEPT:

A: activation of phospholipase C.
B: hydrolysis of cGMP by a phosphodiesterase.
C: liberation of intercellular calcium.
D: opening of potassium channels.
E: production of cAMP by adenylate cyclase.

Learning Response C: Correct. The activation of various G proteins by their receptors (or related rhodopsin) leads to direct activation of adenylate cyclase, cGMP phosphodiesterase, potassium channels, cell membrane calcium channels, and phospholipase C, which produces second messengers which, in turn activate protein kinase C and liberate intercellular calcium.

---

269. SUBJECT AREA: Cell Signaling
QUESTION: Which of the following accurately describes cellular calcium?

A: energy is required to pump it into the cell.
B: it is evenly distributed throughout the cell.
C: it is sequestered into the rough endoplasmic reticulum.
D: its concentration is higher than outside the cell.
E: its transport is sometimes coupled with sodium.

Learning Response E: Correct. Intercellular calcium is kept lower in the cytoplasm through its ATP- or sodium-dependent transport out of the cell, or into intercellular storage compartments which are distinct from the RER.

---

270. SUBJECT AREA: Cell Signaling
QUESTION: The action of phospholipase C:

A: causes calcium sequestration.
B: is as a serine kinase.
C: is stimulated by G_s.
D: produces inositol trisphosphate.

E: uses diacylglycerol as substrate.

Learning Response D: Correct. The activation of phospholipase C by G_p leads to the cleavage of PIP_2 into diacylglycerol and IP_3, which in turn activate protein kinase C and liberate calcium.

---

271. SUBJECT AREA: Cell Signaling
QUESTION: All of the following will result in a stimulation of protein kinase C EXCEPT:

A: calcium ionophores.
B: diacylglycerol.
C: inositol trisphosphate.
D: phorbol esters.
E: all if the above will activate protein kinase C.

Learning Response E: Correct. Protein kinase C, a serine/threonine kinase, is activated by calcium. Thus both IP_3 and calcium ionophores can be used to stimulate it. In addition, diacylglycerol (and phorbol esters, which mimic the activity of this second messenger) increases the sensitivity of the enzyme for calcium, and thus accomplishes the same end.

---

272. SUBJECT AREA: Cell Signaling
QUESTION: Protein kinase A:

A: binds calmodulin.
B: can regulate the calcium signal.
C: exists as a dimer, with a regulatory and a catalytic subunit.
D: is a tyrosine kinase.
E: is inactivated by cAMP.

Learning Response B: Correct. Protein kinase A is a heterotetramer, with two regulatory and two catalytic subunits. The binding of ATP releases the catalytic subunits, which then phosphorylates serines or threonines on numerous proteins. Possible targets include calcium channels, calcium calmodulin kinases, and targets of these kinases.

273. Subject Area: Cell Signaling
Question: The binding of calcium to calmodulin:

I. activates it as a kinase.
II. alters its binding characteristics for other proteins.
III. involves multiple calcium binding sites.
IV. is relatively non-specific.

A: I and II.
B: II and III.
C: II and IV.
D: III and IV.
E: II, III, and IV.

Learning Response B: Correct. Calmodulin contains four highly specific calcium binding sites. When these are occupied, a conformational change in the protein enables it to bind to calcium calmodulin dependent kinases and other sensitive proteins.

---

274. Subject Area: Cell Signaling
Question: Which of the following is true about guanylate cyclase production of cGMP?

A: it is activated by transducin.
B: the cGMP also activates kinase A.
C: the cGMP can act by direct binding to an ion channel.
D: the cGMP works through mechanisms entirely different than cAMP.
E: the enzyme only exists as a soluble cytosolic enzyme.

Learning Response C: Correct. Guanylate cyclase exists in soluble and transmembrane forms. The cGMP that it produces acts through analogous mechanisms to cAMP, but binds distinct kinases and channels specific to it. The action of guanylate cyclase is offset by transducin, which stimulates a cGMP phosphodiesterase.

275. SUBJECT AREA: Cell Signaling

QUESTION: The time required to change the level of a particular protein to a new concentration:

I. depends on the final concentration that is to be reached.
II. depends on the rate at which the protein is being eliminated.
III. depends on the rate at which the protein is being synthesized.

A: I.
B: II.
C: III.
D: I and III.
E: II and III.

Learning Response B: Correct. The amount of time necessary to go from any steady state level to any other steady state level, independent of both the magnitude and direction of the change, is a function of the half-life of elimination of the protein. Thus a protein with a short half life can change quickly, while one with a long half life will take longer.

---

276. SUBJECT AREA: Cell Signaling

QUESTION: All of the following are examples of strategies whereby cells use cooperative approaches to cause a sharper onset of effect, EXCEPT:

A: accelerating the rate of production of hormone, while slowing its elimination.
B: positive feedback loop, with cells signaling back to increase the signal to which they are responding.
C: requirement for dimerization of DNA binding proteins before they can bind DNA.
D: requirement for multiple binding sites to be occupied before any activity is achieved.
E: stimulation of one enzyme and inhibition of the enzyme with the opposing effect.

Learning Response A: Correct. Were one to decrease the rate of degradation of a hormone, the result would be a more gradual onset. This may result in a different maximal stimulation, but would not produce a sharper rate of onset through cooperativity.

---

277. SUBJECT AREA: Cell Signaling
QUESTION: All of the following are mechanisms used by the cell to reduce the response to a hormone signal, EXCEPT:

A: alteration of G protein.
B: internalization of receptors.
C: phosphorylation of receptor.
D: reduction of half-life of receptor.
E: all of the above are used by some cells.

Learning Response E: Correct. While the reduction in the number of available receptors, by their degradation or sequestration, can reduce the response of a cell, another strategy involves the inhibition of the ability of a receptor to pass on its signal, through covalent modification of either the G protein or the receptor.

---

278. SUBJECT AREA: Cell Signaling
QUESTION: Which of the following accurately describe bacterial flagella?

A: it always rotates in the same direction.
B: it is similar in structure to eucaryotic flagella.
C: its motion accounts for the movement seen in chemotaxis.
D: its movement is driven directly by ATP.
E: its speed of rotation is regulated by chemotactic stimuli.

Learning Response C: Correct. Bacterial flagella are entirely different in structure and function from eucaryotic flagella. They are driven by the proton gradient of the cells, and switch back and forth between clockwise and counterclockwise motion. Chemotaxis is brought about through the regulation of the direction of rotation of the flagella.

279. SUBJECT AREA: Cell Signaling
QUESTION: Bacterial chemotactic receptors:

A: have different structures.
B: also transport sugars into the cell.
C: are covalently modified to bring about adaptation.
D: directly bind dipeptides to initiate their signal.
E: must be methylated in order to bind their ligand.

Learning Response C: Correct. These receptors, which all have similar structures, either bind their ligand directly (amino acids) or bind a complex of the attractant and a periplasmic binding protein (dipeptides and sugars). After binding, they are slowly methylated to bring about adaptation.

---

280. SUBJECT AREA: Cell Signaling
QUESTION: The transmission of the chemotactic receptor signal to the flagella:

A: has no feedback to the receptors.
B: can only be studied *in vivo*.
C: involves a cascade of second messengers.
D: involves CheY activation of smooth swimming.
E: requires ATP.

Learning Response E: Correct. The process of transmission of the chemotactic signal from the receptor to the flagella has been studied *in vitro*, and been shown to involve a cascade of phosphorylation events which includes feedback mechanisms. The phosphorylation of CheY, the last protein in the sequence, results in its stimulation of the clockwise rotation of the flagella, and hence tumbling.

---

281. SUBJECT AREA: Cell Signaling
QUESTION: Which of the following is a characteristic of the enzyme adenylate cyclase?

A: it can mediate the signaling of many different hormones.

B: it catalyzes the production of cAMP from AMP.
C: it is a soluble cytoplasmic enzymatic activity.
D: it is activated, in mammals, by the *ras* oncogene.
E: it is one type of catalytic cell-surface receptor.

Learning Response A: Correct. Adenylate cyclase is a cell membrane associated catalytic activity that is regulated, via heterotrimeric G proteins, by many different cell surface receptors, to convert ATP into cAMP. It thus serves to initiate a second messenger response to a number of hormones.

---

282. SUBJECT AREA: Cell Growth and Division
QUESTION: Which of the following is not a phase of the mitotic cycle?

A: G0.
B: G1.
C: G2.
D: M.
E: S.

Learning Response A: Correct. The normal mitotic cycle consists of G1 (gap), S (synthesis), G2 (gap), and finally M (mitosis). G0 is a quiescent state into which cells are often diverted when they cease dividing.

---

283. SUBJECT AREA: Cell Growth and Division
QUESTION: The following are true of cells in culture EXCEPT:

A: they can be grown to large populations.
B: they can be held at a particular stage by use of a drug.
C: they can be synchronized through specific manipulations.
D: they round up from the plate during mitosis.
E: they spontaneously synchronize under certain conditions.

Learning Response E: Correct. Through the use of tissue culture, large populations of synchronous cells can be prepared. While this does not occur spontaneously, cells can be synchronized by one of two ways. Their progress can be blocked at a specific stage by a drug, and then drug removal will allow the cells to proceed together. Alternatively, it is easy to remove cells which have rounded up in mitosis and plate them as a separate synchronous culture.

---

284. SUBJECT AREA: Cell Growth and Division
QUESTION: The following are true of the S phase EXCEPT:

A: cells will not leave S phase until the DNA has been completely replicated.
B: it can be studied through fusion with cells in other stages.
C: it presents an opportunity to label a cell with radioactive nucleotides.
D: once the DNA has been replicated a block prevents further replication.
E: replication is initiated solely by nuclear events.

Learning Response E: Correct. The use of cell fusions and radiolabeling of DNA has enabled the study of controls on S phase. A cytoplasmic signal initiates replication which occurs a single time and then is blocked from recurring until after mitosis.

---

285. SUBJECT AREA: Cell Growth and Division
QUESTION: All the following are true of M-phase-promoting factor EXCEPT:

A: it can be injected into cells.
B: it is produced in an oscillating pattern.
C: it is produced in the nucleus.
D: it promotes entry into M phase.
E: it was discovered in an unfertilized egg.

Learning Response C: Correct. MPF is the factor isolated from mature eggs which enables them to divide rapidly without significant growth phases following fertilization. It is produced in oscillating manner in the cytoplasm and when injected into oocytes will cause them to enter mitosis.

---

286. SUBJECT AREA: Cell Growth and Division
QUESTION: Yeast are convenient for the study of cell division for the following reasons EXCEPT:

A: a diploid cell can conjugate with other diploid cells.
B: it is easy to characterize mutants in yeast.
C: they can proliferate either in diploid or haploid states.
D: they demonstrate two different patterns of division.
E: they reproduce as rapidly as bacteria.

Learning Response A: Correct. The primary advantage for studying cell division in yeast deals with their easy manipulation. They can be grown rapidly to large numbers and mutations can be clonally selected. They exist as either diploid or haploid and the haploid cells can be mated. This allows mutants to be characterized as single copies or combined with a good copy to determine dominance. They also demonstrate both budding and fission.

---

287. SUBJECT AREA: Cell Growth and Division
QUESTION: The following are true of Start EXCEPT:

A: a critical size is necessary for a cell to pass through Start.
B: a protein kinase is necessary for a cell to pass through Start.
C: it is equivalent to the mammalian restriction point.
D: it is the point after which the progress through the cycle can no longer be regulated.
E: it occurs at different times in the cytoplasmic, chromosome and centrosome cycles.

Learning Response E: Correct. Start (or the mammalian restriction points) marks the transition from G1 to S phase. It initiates the process of cell division and requires cells to be of critical size and for a specific protein kinase to be active.

288. SUBJECT AREA: Cell Growth and Division

QUESTION: Which of the following statements about the centrosome, chromosome, and cytoplasmic cycles is accurate?

A: the chromosome cycle is the first to initiate.
B: the chromosome involves DNA synthesis and mitosis.
C: the cytoplasmic cycle alternates between cytokinesis and mitosis.
D: the cytoplasmic cycle is the first to be completed.
E: they are three sequential divisions of the cell cycle.

Learning Response B: Correct. It is sometimes convenient to view cell division in terms of these three parallel cycles. They involve, respectively, replication of the centromeres, synthesis and mitosis of DNA, and growth and cytokinesis. The centrosome cycle begins first, while the cycles are terminated by cytokinesis.

289. SUBJECT AREA: Cell Growth and Division

QUESTION: Which of the following describes the role of growth factors in cell growth?

A: a single growth factor will only stimulate one cell type.
B: each cell requires a single growth factor to proliferate.
C: growth factors are competed for by adjacent cells.
D: growth factors bind the same cellular receptor.
E: they are peripherally involved in regulating cell division.

Learning Response C: Correct. Growth factors are present in small amounts and are competed for by neighboring cells. Cells can be activated by multiple growth factors and each growth factor can activate different cell types. However, each growth factor has a specific receptor to which it binds.

290. SUBJECT AREA: Cell Growth and Division
QUESTION: All the following negatively impact a cell's ability to divide EXCEPT:

A: entrance into G0.
B: loss of anchorage to a surface.
C: removal of adjacent cells.
D: scarcity of growth factor.
E: the number of times that a cell has divided.

Learning Response C: Correct. Removal of adjacent cells will release contact inhibition and promote cell division. With increasing cell divisions, loss of anchorage or reduced availability of growth factor cells will be more likely to enter G0.

---

291. SUBJECT AREA: Cell Growth and Division
QUESTION: Mammalian cell division:

A: does not occur in mature tissues.
B: involves fixed times for each phase.
C: is usually arrested prior to M phase.
D: occurs at the same rate in all cells.
E: requires passage through a restriction point.

Learning Response E: Correct. Division in mammalian cells occurs at different rates in different tissues and even between cells of the same type. Cells spend a variable length of time paused in G0 before passing the restriction point and proceeding into S phase.

---

292. SUBJECT AREA: Cell Growth and Division
QUESTION: All the following sometimes characterize transformed cells EXCEPT:

A: they become anchorage independent.
B: they become contact inhibited.
C: they do not undergo senescence.
D: they pile on top of each other.

E: they proliferate without growth factors.

Learning Response B: Correct. Transformed cells typically lose the characteristics of normal cells that control division. They lose contact inhibition, anchorage and growth factor dependence and senescence. They become immortal and can cause tumors in susceptible animals.

---

293. SUBJECT AREA: Cell Growth and Division
QUESTION: Which of the following is true of tumor causing retroviruses?

A: carry genes with no cellular counterparts.
B: carry oncogenes which control their reproduction.
C: have DNA genomes.
D: have only recently been identified.
E: have provided an easy source of cell cycle regulators.

Learning Response E: Correct. Retroviruses provided a first available means of identifying cellular genes involved in cell regulation. Having an RNA genome they carry a superfluous oncogene. These oncogenes have been cloned and their cellular proto-oncogenes identified.

---

294. SUBJECT AREA: Cell Growth and Division
QUESTION: Oncogenes have allowed the identification of a role in cell division for which of the following?

I. DNA binding proteins.
II. growth factor receptors.
III. signal transducers
IV. threonine kinases.

A: I, II and III.
B: I, II and IV.
C: I, III and IV.
D: II, III and IV.
E: I, II, III and IV.

Learning Response E: Correct. All these have been identified as regulators of cell division, as well as growth factors, tyrosine kinases, and G proteins.

---

295. SUBJECT AREA: Cell Growth and Division
QUESTION: In normal cells the loss of adhesion during mitosis probably involves all the following EXCEPT:

A: activation of a growth factor receptor.
B: activation of *src* kinase.
C: phosphorylation of fibronectin.
D: phosphorylation of v-*src*.
E: secretion of plasminogen activator.

Learning Response D: Correct. The effects of v-src on fibronectin and plasminogen activator are thought to mimic what happens in normal cells. However, in normal cells activation of a growth factor receptor would lead to phosphorylation of c-src and the activation of its kinase activity.

---

296. SUBJECT AREA: Cell Growth and Division
QUESTION: Which stage of mitosis involves the separation of chromosomes?

A: anaphase.
B: metaphase.
C: prometaphase.
D: prophase.
E: telophase.

Learning Response A: Correct. Following alignment of the chromosomes during metaphase, the separation of the kinetochores in anaphase result in the chromosomes being pulled toward opposite poles.

---

297. SUBJECT AREA: Cell Growth and Division
QUESTION: Which stage of mitosis is the longest?

A: anaphase.
B: metaphase.

C: prometaphase.
D: prophase.
E: telophase.

Learning Response B: Correct. It is the alignment of the chromosomes in metaphase that usually takes the longest time while their separation in anaphase takes the shortest.

---

298. SUBJECT AREA: Cell Growth and Division
QUESTION: Which of the following contains actin filaments?

A: aster.
B: centriole.
C: contractile ring.
D: kinetochore.
E: mitotic spindle.

Learning Response C: Correct. During cytokinesis, actin and myosin form a contractile ring which draws the plasma membrane inward between the reforming nuclei. The cleavage furrow deepens until the two sides separate to form the daughter cells.

---

299. SUBJECT AREA: Cell Growth and Division
QUESTION: The kinetochore:

A: contracts and lengthens through microtubule growth.
B: forms a ring to separate the daughter cells.
C: is the center of the aster.
D: is the site of chromosome attachment to the mitotic spindle.
E: serves to nucleate the reforming nuclear envelope.

Learning Response D: Correct. Kinetochores capture the plus ends of spindle microtubules attaching them to the centromere region of the chromosomes. The dynamic changes in the microtubules will align the chromosomes and separate them by pulling the kinetochores in opposite directions.

300. SUBJECT AREA: Cell Growth and Division
QUESTION: Anaphase A involves:

A: alignment of the chromosomes.
B: disassembly of tubulin at the kinetochore.
C: elongation of polar microtubules.
D: lengthening of kinetochore microtubules.
E: separation of spindle poles.

Learning Response B: Correct. In anaphase A, disassembly of kinetochore microtubules leads to their shortening and the separation of the chromosomes. This is followed by anaphase B in which the spindle poles move farther apart and the polar microtubules are lengthened.

301. SUBJECT AREA: Cell Growth and Division
QUESTION: Metaphase chromosomes are held at the metaphase plate by:

A: association of their kinetochores.
B: balanced forces of opposing microtubules.
C: direct interaction of the chromosomes.
D: presence of the contractile ring.
E: the remnants of the nuclear envelope.

Learning Response B: Correct. In prometaphase the differential forces on chromosome kinetochores are thought to cause their motion toward the metaphase plate where they are then held in metaphase by the balanced effect of kinetochore microtubules from opposing poles.

302. SUBJECT AREA: Cell Growth and Division
QUESTION: Reassembly of the nuclear envelope involves all the following EXCEPT:

A: dephosphorylation of nuclear lamins.
B: formation around each individual chromosome.
C: reassociation of the nuclear lamina.
D: re-formation of nuclear pores.

E: synthesis of new membranes.

Learning Response E: Correct. In telophase, the nuclear envelope begins to reassemble around each chromosome from the nuclear envelope vesicles created at the time of its dissociation. These associate into a single nucleus where dephosphorylated lamins regenerate the nuclear lamina and nuclear pores are reformed.

---

303. SUBJECT AREA: Cell Growth and Division
QUESTION: All of the following are true of the contractile ring EXCEPT:

A: evidence of it can first be seen in anaphase.
B: it causes formation of the midbody.
C: it continues to contract until the cells separate.
D: it forms the cleavage furrow.
E: it involves the interaction of actin and myosin.

Learning Response C: Correct. The contractile ring forms from actin and myosin in anaphase forming the cleavage furrow. It contracts, partially by the removal of subunits, until it forms the narrow midbody tethering the two cells. It then dissociates.

---

304. SUBJECT AREA: Cell Growth and Division
QUESTION: What distinguishes the mitotic process in higher organisms?

A: condensation of the chromosomes.
B: dissociation of the nuclear envelope.
C: the presence of centrioles.
D: the presence of a mitotic spindle.
E: the use of microtubules.

Learning Response B: Correct. In lower organisms, mitotic processes which probably served as precursors of those seen in higher organisms can be observed. These include the presence of centrioles and mitotic spindles as well as opposing motion along microtubules. What distinguishes the process in higher organisms is the dissolution of the nuclear envelope.

305. SUBJECT AREA: Cell Growth and Division

QUESTION: All of the following are true of cytokinesis in higher plants EXCEPT:

A: it can be dissociated from mitosis.
B: it involves the action of a cleavage furrow.
C: it involves the action of actin.
D: it involves formation of a cell plate.
E: it involves formation of a phragmoplast.

Learning Response B: Correct. Cytokinesis in plants differs from that seen in animals. Rather than starting at the cell surface with a cleavage furrow, it involves formation of a cell plate at the phragmoplast, the equivalent of the animal midbody. As actin pulls the phragmoplast microtubules toward the periphery, this cell plate extends to the existing cell wall.

---

306. SUBJECT AREA: Cell Adhesion, Cell Junctions, and the Extracellular Matrix

QUESTION: All the following are examples of anchoring junctions EXCEPT:

A: adhesion belts
B: desmosomes.
C: focal contacts.
D: hemidesmosomes.
E: tight junctions.

Learning Response E: Correct. Unlike occluding junctions such as tight junctions, anchoring junctions serve to attach cells to their neighbors or to the matrix. They include adherens junctions such as adhesion belts and focal contacts, as well as desmosomes.

---

307. SUBJECT AREA: Cell Adhesion, Cell Junctions, and the Extracellular Matrix

QUESTION: The following are true of tight junctions EXCEPT:

A: they allow diffusion along the membrane.

B: they are often found between epithelial cells.
C: they can be observed by freeze-fracture electron microscopy.
D: they prevent diffusion between cells.
E: they require extracellular calcium.

Learning Response A: Correct. In epithelial cells, tight junctions form a barrier to diffusion both between cells and within the cellular membrane, which action requires extracellular calcium and can be readily observed by freeze-fracture technique.

---

308. SUBJECT AREA: Cell Adhesion, Cell Junctions, and the Extracellular Matrix
QUESTION: Which of the following is true of adherens junctions?

A: they are made up of two functional protein classes.
B: they are only used to link cells together.
C: they involve desmosomes.
D: they link to intermediate filaments.
E: they prevent diffusion through the membrane.

Learning Response A: Correct. Adherens junctions link active filaments between cells or to the extracellular matrix. This is done through the interaction of an extracellular protein and a transmembrane linker glycoprotein.

---

309. SUBJECT AREA: Cell Adhesion, Cell Junctions, and the Extracellular Matrix
QUESTION: Which of the following is true of gap junctions?

A: their ability to form is species dependent.
B: they are limited to specific animal tissues.
C: they can be observed by electron microscopy.
D: they exist as protein monomers.
E: they prevent the passage of molecules of any size.

Learning Response C: Correct. The formation of gap junctions involves the interaction of six identical subunits, forming a pore that permits passage of small molecules between adjacent cells, even if they are from different species. They can be observed in most tissues by electron microscopy.

---

310. SUBJECT AREA: Cell Adhesion, Cell Junctions, and the Extracellular Matrix

QUESTION: Formation and maintenance of which of the following types of cell junctions can be regulated by calcium?

I. anchoring junctions.
II. communicating junctions.
III. tight junctions.

A: I.
B: III.
C: I and II.
D: II and III.
E: I, II and III.

Learning Response E: Correct. Extracellular calcium is necessary for the maintenance of tight junctions. Adherence junctions, a form of anchoring junctions, hold membranes together by a calcium dependent mechanism. In gap junctions high intercellular calcium decreases permeability.

---

311. SUBJECT AREA: Cell Adhesion, Cell Junctions, and the Extracellular Matrix

QUESTION: All of the following are true of desmosomes EXCEPT:

A: they are biochemically distinct in different tissues.
B: they are composed of intercellular and transmembrane regions.
C: they attach to the same type of intermediate filaments.
D: they can be involved in autoimmune disease.
E: they connect intermediate filaments of adjacent cells.

Learning Response C: Correct. Desmosomes link intermediate filaments between adjacent cells through the combined action of intercellular attachment proteins and transmembrane linker glycoproteins. Antibodies from autoimmune patients will disrupt desmosomes in epithelial cells but not in other tissues indicating that they are biochemically distinct and explaining their association with different types of filaments.

---

312. SUBJECT AREA: Cell Adhesion, Cell Junctions, and the Extracellular Matrix

QUESTION: All the following are true of the macromolecules of the extracellular matrix EXCEPT:

A: they are largely secreted by fibroblasts.
B: they are secreted locally.
C: they are free of proteins.
D: they can be divided into two classes.
E: they form a gel-like substance.

Learning Response C: Correct. In most tissues the extracellular matrix is secreted locally by fibroblasts. It is made up of glycosaminoglycans, which form a gel-like ground substance, and fibrous proteins.

---

313. SUBJECT AREA: Cell Adhesion, Cell Junctions, and the Extracellular Matrix

QUESTION: Which of the following is true of glycosaminoglycans?

A: they are extremely flexible.
B: they are formed from branched polysaccharide chains.
C: they are made up of repeating monosaccharides.
D: they differ in their sugars, linkages and sulfate groups.
E: they have hydrophobic tendencies.

Learning Response D: Correct. Glycosaminoglycans are extremely hydrophilic, inflexible and unbranched polysaccharides. They are distinguished by the linkage, sulfation and specific sugars found in their disaccharides repeats.

314. SUBJECT AREA: Cell Adhesion, Cell Junctions, and the Extracellular Matrix

QUESTION: Which of the following glycosaminoglycans is not linked to proteins?

A: chondroitin sulfate.
B: dermatan sulfate.
C: heparin.
D: hyaluronic acid.
E: keratan sulfate.

Learning Response D: Correct. All of these glycosaminoglycans are linked to protein except hyaluronic acid, which is thought to play a role in cell migration.

---

315. SUBJECT AREA: Cell Adhesion, Cell Junctions, and the Extracellular Matrix

QUESTION: Proteoglycans and glycoproteins are usually distinguishable by which of the following?

I. their amino acid composition.
II. their carbohydrate to protein ratio.
III. the length of their oligosaccharide chains.
IV. their total size.

A: I, II and III.
B: I, II and IV.
C: I, III and IV.
D: II, III and IV.
E: I, II, III and IV.

Learning Response D: Correct. While they can contain the same protein core, proteoglycans are usually much larger, have longer oligosaccharide chains and a higher percentage of carbohydrates than do glycoproteins.

---

316. SUBJECT AREA: Cell Adhesion, Cell Junctions, and the Extracellular Matrix

QUESTION: The following are true of proteoglycans in tissues EXCEPT:

A: they are structurally heterogeneous.
B: they associate noncovalently with glycosaminoglycans.
C: they can be easily observed by electron microscopy.
D: they can be integral plasma membrane components.
E: they provide a molecular sieve.

Learning Response C: Correct. Proteoglycans are highly heterogeneous. Either as secreted units or extracellular domains of transmembrane proteins, they interact with glycosaminoglycans to form a sieve-like gel.When isolated from tissue they can easily be studied by electron microscopy but this is much more difficult in their native tissues due to their high degree of hydration and their high solubility.

---

317. SUBJECT AREA: Cell Adhesion, Cell Junctions, and the Extracellular Matrix

QUESTION: Collagen fibrils are formed from all of the following types of collagen EXCEPT:

A: type I.
B: type II.
C: type III.
D: type IV.
E: all of the above are involved in fibril formation.

Learning Response D: Correct. Collagen types I, II and III form fibrils while type IV is involved in the formation of the basal lamina.

---

318. SUBJECT AREA: Cell Adhesion, Cell Junctions, and the Extracellular Matrix

QUESTION: The formation of mature collagen involves all the following EXCEPT:

A: amino acid chains have a repeating sequence motif.
B: formation of a left handed helix by each peptide chain.
C: formation of hydroxyproline.
D: interaction of four collagen helices.
E: proteolytic cleavage.

Learning Response D: Correct. Collagen is comprised of three helical α chains with a repeating G-X-Y sequence. Selected prolines are hydroxylated in the endoplasmic reticulum. Following formation of the appropriate helix pro-peptides are cleaved from both ends.

---

319. SUBJECT AREA: Cell Adhesion, Cell Junctions, and the Extracellular Matrix

QUESTION: Type IV collagen differs from types I, II and III by all the following EXCEPT:

A: they are not cleaved after secretion.
B: they do not form into fibrils.
C: they do not form triple helices.
D: their G-X-Y are locally interrupted.
E: they have a different gene structure.

Learning Response C: Correct. Type IV collagen, which forms a sheet like network rather than fibrils, is still composed of triple helical alignments of individual proteins. This structure is disrupted at irregularities in the G-X-Y repeat, and at the ends of the molecule which are not cleaved. These structural differences are also reflected in the gene structure.

---

320. SUBJECT AREA: Cell Adhesion, Cell Junctions, and the Extracellular Matrix

QUESTION: The following are true of fibronectin EXCEPT:

A: it can form oligomers.
B: it consists of repeats of three types of sequences.
C: it exists as a trimer.
D: it is linked by disulfide bonding

E: it plays a role in cell migration.

Learning Response C: Correct. Fibronectin is formed by the disulfide linkage of two similar proteins made up of repeats of three short sequences. It can exist as soluble dimers, oligomers or insoluble fibrils and functions not only in cell adhesion but in migration.

---

321. SUBJECT AREA: Cell Adhesion, Cell Junctions, and the Extracellular Matrix

QUESTION: The basal lamina is composed of which of the following ?

I. fibronectin.
II. laminin.
III. proteoglycans.
IV. type IV collagen.

A: I, II and III.
B: I, II and IV.
C: I, III and IV.
D: II, III and IV.
E: I, II, III and IV.

Learning Response D: Correct. The basal lamina consists of a mat of type IV collagen with other associated proteins which vary among tissues but always include proteoglycans, laminin and inactin.

---

322. SUBJECT AREA: Cell Adhesion, Cell Junctions, and the Extracellular Matrix

QUESTION: All the following are true of integrins EXCEPT:

A: they are transmembrane proteins.
B: they have similar dimer structures.
C: they help link cells to extracellular matrix.
D: they include all matrix receptors.
E: they share a common β chain.

Learning Response D: Correct. The integrin super family includes only those matrix receptors, such as the fibronectin receptor, that share the same conserve structure. They are transmembrane proteins with a common β subunit and unique α subunits.

---

323. SUBJECT AREA: Cell Adhesion, Cell Junctions, and the Extracellular Matrix

QUESTION: A slime mold will migrate in response to all of the following EXCEPT:

A: cyclic AMP.
B: light.
C: heat.
D: starvation.
E: all of the above.

Learning Response E: Correct. Under conditions of starvation, a cellular slime mold will aggregate through cyclic AMP driven chemotaxis. They form a slug which will then migrate in search of a new food source, toward light or in response to heat.

---

324. SUBJECT AREA: Cell Adhesion, Cell Junctions, and the Extracellular Matrix

QUESTION: The following are true of slime mold cell interaction EXCEPT:

A: it involves a single pass transmembrane protein.
B: it is calcium independent.
C: it is initiated by chemotaxis.
D: it is known to involve binding to an extracellular linker molecule.
E: they appear to involve proteins unrelated to those seen in mammals.

Learning Response D: Correct. Following aggregation mediated by chemotaxis, slime mold cells form contact sites of two types. They first form calcium dependent contact site B interactions and later calcium independent contact site A interactions. Contact site A is a single transmembrane protein which interacts by homophilic binding and is unrelated to any known vertebrate-cell adhesion protein.

---

325. SUBJECT AREA: Cell Adhesion, Cell Junctions, and the Extracellular Matrix

QUESTION: All of the following are true of the neural cell adhesion molecule EXCEPT:

A: it functions in a calcium independent manner.
B: it is believed to work by homophilic interaction.
C: its sequence is distinct from all other known proteins.
D: its structure includes five specially folded domains.
E: it was identified through an immunological strategy.

Learning Response C: Correct. N-CAM was identified immunologically and shown to utilize homophilic interaction to bring about adhesion in a calcium independent manner. Its structure consists of five immunoglobulin-like domains.

---

326. SUBJECT AREA: Cell Adhesion, Cell Junctions, and the Extracellular Matrix

QUESTION: Which of the following statements about cadherins is accurate?

A: all cells express one of the three known classes.
B: they are involved in calcium independent adhesion.
C: they are multi-pass transmembrane proteins.
D: they have immunoglobulin-like domains.
E: they share some characteristics with N-CAM.

Learning Response E: Correct. While sharing some similarities with N-CAM (i.e. they are single pass transmembrane proteins), they are distinct in several ways. They work by a calcium dependent mechanism and do not have immunoglobulin-like structure. It is likely that some have yet to be characterized since the known forms are not found on all cells which experience calcium dependent adhesion.

327. SUBJECT AREA: Cell Adhesion, Cell Junctions, and the Extracellular Matrix

QUESTION: Adhesion in slime mold uses which of the following mechanisms that are observed in vertebrates?

I. chemotaxis.
II. division from founder cells.
III. pathway guidance.

A: I and II.
B: I and III.
C: II and III.
D: I, II and III.
E: none of the above.

Learning Response B: Correct. While cell division does not appear to play a role in slime mold cell-cell adhesion, they do demonstrate chemotaxis in response to cyclic AMP and pathway guidance using discoidin-1.

328. SUBJECT AREA: Germ Cells and Fertilization

QUESTION: A secondary spermatocyte has:

A: 23 chromosomes, 23 chromatids.
B: completed meiosis.
C: the ability to synthesize RNA.
D: undergone one unequal division.
E: undergone spermiogenesis.

Learning Response C: Correct. Secondary spermatocytes represent and intermediate in the process of spermatogenesis. They have undergone one round of equal meiosis, but still have a second round remaining, and thus have 23 chromosomes, with a pair of chromatids for each (46 total). They retain the ability to synthesize RNA.

329. SUBJECT AREA: Germ Cells and Fertilization

QUESTION: In multicellular organisms, sexual reproduction involves all the following EXCEPT:

A: alternative haploid and diploid generations.
B: both mitosis and meiosis.
C: many somatic cells not involved in reproduction.
D: short diploid and long haploid phases.
E: the fusion of gametes.

Learning Response D: Correct. Sexual reproduction in multicellular organisms involves meiosis of germ line cells to produce haploid gametes. These gametes fuse and then undergo mitosis producing somatic cells in addition to the germ cells which will be involved in the next cycle. The haploid phase is thus limited while the diploid is long lived.

---

330. SUBJECT AREA: Germ Cells and Fertilization
QUESTION: Sexual reproduction is advantageous:

A: because it allows for more genetic recombination.
B: because it is simpler than asexual reproduction.
C: because it requires less resources than asexual reproduction.
D: only in an invariant environment.
E: only under unusual conditions.

Learning Response A: Correct. While it is complex and requires more resources under typical conditions of changing environment, the ability to undergo genetic recombination allows cells to adapt more readily.

---

331. SUBJECT AREA: Germ Cells and Fertilization
QUESTION: The advantage of the diploid state to evolution is:

A: it increases the mutation rate.
B: it prevents mutations from occurring.
C: it prevents sexual reproduction.
D: it provides a spare copy of genes.
E: it provides no advantage.

Learning Response D: Correct. While it does not affect the rate at which mutations occur, the presence of a spare gene allows more experimentation with mutations without risking the survival of the organism. When combined with sexual reproduction, it also allows for recombination.

---

332. SUBJECT AREA: Germ Cells and Fertilization
QUESTION: The following are true of gene duplication EXCEPT:

A: it allows for divergence of protein function.
B: it allows for more elaborate genetic recombination.
C: it can be passed to successive generations.
D: it is in part responsible for protein families.
E: it prevents the appearance of mutations.

Learning Response E: Correct. The duplication of genes when coupled to the mutational process allows for two different versions of a sequence to be passed to progeny. This results in the production of families of related genes which in turn can recombine to reveal even more possible sequences thus accelerating the evolution of proteins with novel characteristics.

---

333. SUBJECT AREA: Germ Cells and Fertilization
QUESTION: Meiosis differs from mitosis in all the following EXCEPT:

A: it involves an initial DNA replication step.
B: it involves the alignment of homologous chromosomes.
C: it places higher demands on the division machinery.
D: it produces haploid daughters.
E: it requires two nuclear divisions.

Learning Response A: Correct. Meiosis is more complex and hence is more demanding of the replication process. After an initial round of DNA replication similar to that seen in mitosis there is then an alignment of homologous chromosomes and two rounds of nuclear division producing haploid cells, none of which occurs during mitosis.

334. SUBJECT AREA: Germ Cells and Fertilization
QUESTION: Chromosomal crossing-over involves all the following EXCEPT:

A: formation of bivalent.
B: formation of chiasmata.
C: genetic recombination.
D: takes place during meiotic division II.
E: two sister chromatids.

Learning Response D: Correct. Following formation of a bivalent by sister chromatids in meiosis I, crossing-over at chiasmata allow for genetic recombination.

---

335. SUBJECT AREA: Germ Cells and Fertilization
QUESTION: Which of the following is the correct order of the stages of meiotic prophase I?

A: diakinesis, diplotene, lepotene, pachytene, zygotene.
B: diplotene, zygotene, diakinesis, pachytene, lepotene.
C: lepotene, zygotene, pachytene, diplotene, diakinesis.
D: pachytene, lepotene, diakinesis, zygotene, diplotene.
E: zygotene, lepotene, diplotene, pachytene, diakinesis.

Learning Response C: Correct. The stages of mitotic prophase I as defined by the morphological changes in the DNA are lepotene, zygotene, pachytene, diplotene and diakinesis.

---

336. SUBJECT AREA: Germ Cells and Fertilization
QUESTION: All the following are true of the synaptonemal complex EXCEPT:

A: it consists of a ladder-like protein core.
B: it dissolves just after the pachytene stage.
C: it forms just before the diplotene stage.
D: it is thought to be required for crossing-over.
E: it keeps the homologous chromosomes in a bivalent.

Learning Response C: Correct. The synaptonemal complex exists only during the pachytene stage. Its ladder-like protein core aligns homologous chromosomes and is thought to allow for recombination.

---

337. SUBJECT AREA: Germ Cells and Fertilization
QUESTION: In what manner is it insured that sex chromosomes will segregate?

A: by a specific recognition protein.
B: by conserving small homology between them.
C: by duplicating them.
D: by pairing them with autosomes.
E: they do not segregate.

Learning Response B: Correct. Small regions at the end of the X and Y chromosomes undergo homologous pairing and crossing over and thereby act as homologous chromosomes.

---

338. SUBJECT AREA: Germ Cells and Fertilization
QUESTION: The following are true of recombination nodules EXCEPT:

A: they are thought to be responsible for recombination.
B: they are very small in size.
C: they incorporate radioactive DNA into chromosomes.
D: they share the same chromosomal distribution as crossover events.

E: their total number is about equal to the total number of the chiasmata.

Learning Response B: Correct. These structures are very large and are implicated in crossing over due to a similar distribution to chiasmata and crossover events, and a DNA synthesis activity required for recombination.

---

339. SUBJECT AREA: Germ Cells and Fertilization
QUESTION: The following are true of an egg EXCEPT:

A: it can give rise to every cell type in the adult.
B: it is a highly specialized cell.
C: it has large reserves of RNA.
D: it is larger than somatic cells.
E: it requires fertilization to develop.

Learning Response E: Correct. Eggs are large, highly specialized cells with large reserves of RNA and ribosomes that enable it to divide rapidly and eventually develop into an entire organism. The activation of this process usually involves fertilization but it can be activated in other ways.

---

340. SUBJECT AREA: Germ Cells and Fertilization
QUESTION: The following are characteristics of mammalian eggs EXCEPT:

A: they are 1/10 the diameter of fish eggs.
B: they are devoid of yolk.
C: they contain specialized secretory vesicles.
D: they develop in stages.
E: they have specialized extracellular matrix.

Learning Response B: Correct. Mammalian eggs develop in stages and have a distinct extracellular matrix referred to as the zona pellucida. They differ from eggs of other species in their size, their possession of cortical granules and in being less than 5% yolk.

341. SUBJECT AREA: Germ Cells and Fertilization
QUESTION: Oogenesis involves which of the following?

A: division of oogonia by mitosis.
B: production of four haploid eggs.
C: rapid progression through meiosis.
D: release of the primary oocyte at ovulation.
E: symmetrical division.

Learning Response A: Correct. Oogonia divide repeatedly by mitosis. After differentiating into primary oocytes, they initiate meiosis which is paused at prophase allowing the oocyte to further develop. It then completes two rounds of asymmetric meiosis and produces a single egg to be released at ovulation.

---

342. SUBJECT AREA: Germ Cells and Fertilization
QUESTION: In vertebrates, oocyte maturation involves all of the following EXCEPT:

A: a decrease in intercellular cyclic AMP.
B: activation of M-phase promoting factor.
C: contribution of small molecules from follicle cells.
D: direct response to gonadotropins.
E: progression through meiosis I.

Learning Response D: Correct. Follicle cells play two roles in the maturation of oocytes. In addition to feeding small molecules through gap junctions, they respond to gonadotropins by secreting progesterone which reduces cyclic AMP. This leads to activation of M-phase promoting factor and progression through meiosis I.

---

343. SUBJECT AREA: Germ Cells and Fertilization
QUESTION: The development of a human oocyte involves all the following EXCEPT:

A: an arrest in miotic prophase I for up to 40 years.
B: an arrest in miotic prophase II until fertilization.
C: complete maturation of 15-20 oocytes at a time.

D: degeneration of most follicles.
E: formation of an antrum.

Learning Response C: Correct. Development of an oocyte involves the development of a follicle, which will develop an antrum. Following puberty 10 to 15 primary oocytes will progress toward maturation with all but one digressing. The mature oocyte is arrested in prophase II awaiting fertilization.

---

344. SUBJECT AREA: Germ Cells and Fertilization
QUESTION: Adaptations of sperm for delivering their DNA include all the following EXCEPT:

A: a large secretory vesicle of hydrolytic enzymes.
B: a strong flagellum.
C: large stores of RNA.
D: minimal cytoplasm.
E: specialized mitochondria.

Learning Response C: Correct. Sperm contain an acrosomal vesicle full of enzymes for breaking down the egg's outer coat, a nucleus, specialized mitochondria driving a strong flagellum and little else.

---

345. SUBJECT AREA: Germ Cells and Fertilization
QUESTION: Which of the following could explain the availability of the full diploid genome to sperm development?

I. all information necessary for development is established prior to meiosis II.
II. connection between haploid cells enable each sperm to benefit from all the genes present.
III. they remain diploid after division.

A: II.
B: I and II.
C: I and III.
D: II and III.
E: I, II and III.

Learning Response B: Correct. In *Drosophila*, sperm have been observed to develop without DNA. This process could be explained by communication through the connections known to exist between spermatids, or if all the necessary information was present in the cells prior to division.

---

346. SUBJECT AREA: Germ Cells and Fertilization

QUESTION: The acrosomal process becomes coated with the following from the acrosomal vesicle EXCEPT:

A: acrosomal vesicle membrane.
B: binding proteins specific for the vitelline layer.
C: hydrolytic enzymes that penetrate the jelly coat.
D: hydrolytic enzymes that penetrate the vitelline layer.
E: unpolymerized actin.

Learning Response E: Correct. Polymerization of actin drives the acrosomal process through the remnants of the acrosomal vesicle where in addition to remnants of the vesicle membrane, it also accumulates enzymes which enable it to bind and penetrate the jelly coat and vitelline layer to gain access to the plasma membrane.

---

347. SUBJECT AREA: Germ Cells and Fertilization

QUESTION: The species specific fertilization of externally fertilized eggs is guaranteed by:

A: confinement to a small space.
B: random luck.
C: parthenogenesis.
D: specific hydrolytic enzymes.
E: specific vitelline layer binding proteins.

Learning Response E: Correct. A species specific protein named bindin is thought to mediate both binding to the vitelline layer and fusion of the plasma membranes.

348. SUBJECT AREA: Germ Cells and Fertilization
QUESTION: Egg activation involves which of the following:

A: decrease in pH.
B: efflux of sodium.
C: increase in free calcium.
D: membrane polarization.
E: synthesis of new proteins.

Learning Response C: Correct. Activation can occur in the absence of protein synthesis. It causes a rapid membrane depolarization, a release of calcium stores and a sodium coupled efflux of protons, increasing the pH.

349. SUBJECT AREA: Germ Cells and Fertilization
QUESTION: The calcium increase which allows for egg inactivation:

A: appears to be G protein mediated.
B: can act by binding calmodulin.
C: can be mimicked artificially.
D: is a response to the cell polarization.
E: is blocked by chelators.

Learning Response D: Correct. Fertilization appears to activate a G protein which causes the release in calcium. This is implicated as the activator of eggs, as activation can be mimicked by increasing calcium, and blocked by chelating it. The calcium acts through calmodulin and other pathways to activate the egg.

350. SUBJECT AREA: Germ Cells and Fertilization
QUESTION: The fertilization of mammalian eggs differs from the sea urchin in all of the following ways, EXCEPT:

A: it does not involve an acrosomal process.
B: the pronuclei do not fuse prior to the first mitosis.
C: the sperm does not contain centrioles.
D: there is no acrosomal reaction.

E: there is no fast block.

Learning Response D: Correct. Following the acrosomal reaction, the mammalian sperm membrane fuses with the plasma membrane. Because of the low number of sperm that approach the egg, no fast block is needed. The male pronucleus enters the cell, and the female centrioles direct the first division, without fusion of the pronuclei.

---

351. SUBJECT AREA: Germ Cells and Fertilization
QUESTION: The intercellular pH of fertilized eggs:

A: following an initial spike, is rapidly returned to normal.
B: requires mRNA synthesis to have its effect.
C: is altered by a sodium/proton exchanger.
D: is lowered following activation.
E: represses protein synthesis.

Learning Response C: Correct. Activation of a exchanger causes a rise in intercellular pH. This is maintained throughout zygote development, and results in increased protein synthesis that does not require RNA synthesis.

---

352. SUBJECT AREA: Cellular Mechanisms of Development
QUESTION: Which of the following refers to the axis of an animal progressing from head to tail:

A: animal-vegetal.
B: anterior-posterior.
C: dorso-ventral.
D: medio-lateral.
E: none of the above.

Learning Response B: Correct. Of the three body axes, anterior-posterior refers to head-tail, while dorso-ventral is back to belly, and medio-lateral is center-left/right. An egg demonstrates animal-vegetal polarity, with the yolk toward the vegetal pole.

353. SUBJECT AREA: Cellular Mechanisms of Development
QUESTION: In the sea urchin, migration of primary mesenchyme cells is thought to involve all of the following, EXCEPT:

A: an affinity for fibronectin-rich matrix.
B: invagination of the mesenchymal cells.
C: location of surfaces to which they can firmly attach.
D: loss of their propensity to stick to other cells.
E: use of filopodia to move along the inner wall.

Learning Response B: Correct. Mesenchymal cells appear to no longer stick to other cells but instead prefer the fibronectin of the matrix. They migrate along the surface using filopodia until they locate a region to which they can firmly attach. At the same time, the epithelium invaginates to form the gut.

---

354. SUBJECT AREA: Cellular Mechanisms of Development
QUESTION: Which of the following is correct?

A: the ectoderm gives rise to the notochord.
B: the endoderm gives rise to the liver.
C: the endoderm gives rise to the vasculature.
D: the mesoderm gives rise to the lungs.
E: the mesoderm gives rise to the nervous system.

Learning Response B: Correct. The endoderm produces the gut, liver, lungs and associated organs. From the mesoderm comes muscles, connective tissue, vascular and urogenital, and includes the notochord. The epidermis as well as the nervous system arise from the ectoderm.

---

355. SUBJECT AREA: Cellular Mechanisms of Development
QUESTION: Neurulation involves all the following EXCEPT:

A: cell elongation perpendicular to the epithelial sheet.
B: formation of neural folds.
C: migration of individual cells through the

mesoderm.
D: pinching of the neural tube from the ectoderm.
E: thickening of the endoderm at the midline.

Learning Response E: Correct. In neurulation, the ectoderm thickens along the midline. The lateral edges form folds which seal over, forming the neural tube which eventually pinches off from the epidermis with a number of cells breaking loose and migrating into the mesoderm.

---

356. SUBJECT AREA: Cellular Mechanisms of Development
QUESTION: The following are true of determination EXCEPT:

A: it can be investigated by transplantation.
B: it can involve cytoplasmic or nuclear mechanisms.
C: it commits a cell to a specialized course of development.
D: it is a process which is temporary.
E: it usually occurs prior to differentiation.

Learning Response D: Correct. Determination involves the development of a distinguishable difference in a cell and its progeny. This specialized development has been investigated through transplantation and has been found to be either cytoplasmic or chromosomal in nature.

---

357. SUBJECT AREA: Cellular Mechanisms of Development
QUESTION: Which of the following is NOT true about mammalian embryonic development?

A: compaction occurs to the morula.
B: intercellular spaces enlarge to form the blastocoel.
C: the embryo derives from the inner cell mass.
D: the trophectoderm gives rise to the placenta.
E: tight junctions first form in the blastocyst.

Learning Response E: Correct. The morula begins to compact and form tight junctions after the eight-cell stage. The blastocoel develops from enlargement of intercellular spaces producing the blastocyst. This is made up of trophectoderm lining the blastocoel and a denser inner cell mass, of which the former develops into the placenta while the latter forms the embryo.

---

358. SUBJECT AREA: Cellular Mechanisms of Development
QUESTION: The following are true of imprinting EXCEPT:

A: it can be avoided by using two female pronuclei.
B: it causes the pronuclei to have different characteristics.
C: it involves DNA methylation.
D: it prevents mammalian virgin birth.
E: its pattern is preserved in non-germ-line tissues.

Learning Response A: Correct. Through methylation of DNA, imprinting confers different characteristics to the pronuclei that can be inherited by all progeny cells. This prevents virgin birth as a cell with two copies of female-derived chromosomes fails to properly develop.

---

359. SUBJECT AREA: Cellular Mechanisms of Development
QUESTION: The following are true of cells in the eight-cell morula, EXCEPT:

A: each cell can form any part of the adult.
B: none can be destroyed without affecting development.
C: they can be combined to form a chimera.
D: they can be separated and cloned.
E: they are all totipotent.

Learning Response B: Correct. Through the eight-cell stage, the cells of a mammalian embryo are not committed to any particular developmental course. They can be fused or separated or some destroyed without affecting the development of the embryo.

---

360. SUBJECT AREA: Cellular Mechanisms of Development

QUESTION: The following are advantages to studying development in the nematode system, EXCEPT:

A: it has a defined cell lineage.
B: it has a simple genomic arrangement.
C: it is transparent.
D: it only reproduces asexually.
E: its anatomy is simple.

Learning Response D: Correct. The nematode has most of the structures of a more advanced animal, but simplified genetics and anatomy. The invisible body of the worm has allowed its entire cell lineage to be determined. It can reproduce either sexually or asexually, producing homozygotes from the latter.

---

361. SUBJECT AREA: Cellular Mechanisms of Development
QUESTION: The following are true of nematode development, EXCEPT:

A: differentiation is coupled to division.
B: it can be studied by observing the affects of specific mutations.
C: it is defined by developmental control genes.
D: it is directed primarily by cell-cell interactions.
E: it is nearly invariant.

Learning Response D: Correct. The process of development in differentiation in nematodes is coupled to cell division. It involves developmental control genes which have been studied through their mutation and depends on both cell memory and extrinsic control.

---

362. SUBJECT AREA: Cellular Mechanisms of Development
QUESTION: Which of the following is true of rat glial cell differentiation?

A: it can only be studied *in vivo*.
B: their development is controlled by growth factors from nearby cells.

C: they differentiate into astrocytes in their default pathway in culture.
D: they rely on cell-cell contact to differentiate properly.
E: when separated from their environment, they continue dividing.

Learning Response B: Correct. Rat glial cells can be caused to differentiate in culture in a manner similar to that seen *in vivo* if they are provided with appropriate growth factors. In the absence of these growth factors, they stop dividing and differentiate prematurely.

---

363. SUBJECT AREA: Cellular Mechanisms of Development
QUESTION: All the following are true of the action of a morphogen EXCEPT:

A: it can be responsible for establishment of polarity.
B: it can impose asymmetry on neighboring cells.
C: it often must meet a threshold to have its desired effect.
D: its signal can be enhanced by positive feedback.
E: most have been purified and tested *in vitro*.

Learning Response E: Correct. Morphogens are thought to act through establishing a concentration gradient which can then be amplified by the recipient cells through feedback mechanisms or threshold effects to create asymmetry. Relatively few morphogens have been isolated and studied to date.

---

364. SUBJECT AREA: Cellular Mechanisms of Development
QUESTION: Grafting experiments with chicken limb buds have revealed which of the following?

A: bud tissue is committed early to its final form.
B: development involves both memory and position effects.
C: limb development is entirely dependent on positional effects.

D: transfer of limb tissue prevents its development.
E: transplantation has no effect on development.

Learning Response B: Correct. When primordial thigh tissue is placed at the tip of a wing bud, it develops into toes, indicating that it is already committed to leg development but position effects determine the region of the leg it will become.

---

365. SUBJECT AREA: Cellular Mechanisms of Development
QUESTION: All of the following are true of development of the vertebrate limb bud, EXCEPT:

A: appears to be mediated by retinoic acid.
B: developmental signals are unique to each limb.
C: has been studied by grafting.
D: involves a morphogen gradient.
E: is directed by a polarizing region.

Learning Response B: Correct. Development of the limb bud appears to involve a morphogen gradient produced by secretion of retinoic acid from a polarizing region. The polarizing region from any limb or any species will have the same effect on chicken limb development.

---

366. SUBJECT AREA: Cellular Mechanisms of Development
QUESTION: The following are true of *Drosophila* development, EXCEPT:

A: it begins in the form of a syncytium.
B: it can be affected through specific mutations.
C: it involves modulation of a segmental pattern.
D: it involves three different sets of pattern-control genes.
E: its ground plan is not established until pupation.

Learning Response E: Correct. The ground plan for *Drosophila* development is laid during the first hours after fertilization while it is still forming a syncytium. It leads to a segmental pattern that is governed by three classes of pattern-control genes: egg-polarity gene, segmentation genes and homeotic selector genes. These can be studied by analysis of specific mutation.

---

367. SUBJECT AREA: Cellular Mechanisms of Development
QUESTION: The polarity of *Drosophila* is determined:

A: at the time of fertilization
B: by the formation of the cellular blastoderm.
C: by homeotic selector genes.
D: by the position of gastrulation.
E: prior to fertilization.

Learning Response E: Correct. Embryo polarity is determined by a polarity established in the egg prior to fertilization. It is dependent on the product of egg polarity genes and involves a gradient which can be manipulated experimentally.

---

368. SUBJECT AREA: Cellular Mechanisms of Development
QUESTION: All the following are true of segmentation genes, EXCEPT:

A: several mutant genes have been characterized.
B: their expression is regulated by position effects.
C: their mutation results in loss of adjacent segments.
D: they are divided into three classes.
E: they are zygotic-effect genes.

Learning Response C: Correct. Segmentation genes, which are the first developmental genes whose signals are derived from the zygotic genome, can be divided into three classes: gap genes, pair-rule genes and segment-polarity genes. These have been extensively studied through analysis of mutations and have differing effects on the embryo. These effects are not only seen in the cells expressing the genes but also positioning effects on adjacent regions are observed.

---

369. SUBJECT AREA: Cellular Mechanisms of Development

QUESTION: The following are true of homeobox genes, EXCEPT:

A: their products localize to the cytoplasm.
B: they are involved in mammalian development.
C: they contain a conserved sequence.
D: they control the segmentation process.
E: they have arisen through gene duplication.

Learning Response A: Correct. The development of segmentation in *Drosophila* is controlled by the homeobox genes, which contain a conserved homeobox region. They are DNA binding proteins which appear to have arisen by gene duplication. Related genes play similar roles in mammalian development.

---

370. SUBJECT AREA: Cellular Mechanisms of Development
QUESTION: Which of the following is true of homeotic selector genes?

A: their expression is regulated by anterior gene products.
B: their order of expression is reflected in their chromosomal location.
C: they are expressed in the same region of each segment.
D: they are present transiently.
E: they are turned on after gastrulation.

Learning Response B: Correct. These genes are permanently activated in the blastoderm. Their location on the chromosome governs their expression along the length of the embryo with each gene being repressed by the product of that posterior to it.

---

371. SUBJECT AREA: Cellular Mechanisms of Development
QUESTION: The following are true of imaginal disks, EXCEPT:

A: their developmental course is predetermined.
B: their differentiation depends on homeotic selector genes.

C: they give rise to the entire adult.
D: they originate from the embryonic epidermis.
E: they possess a stable heritable memory.

Learning Response C: Correct. The imaginal disks are derived from the epidermis and give rise to the majority of the adult. Their developmental fate is predetermined and heritable, the result of the expression of homeotic selector genes.

---

372. SUBJECT AREA: Cellular Mechanisms of Development
QUESTION: Which of the following are involved in regulating homeotic gene expression?

I. chromosomal arrangement
II. chromosomal rearrangement
III. RNA splicing.
IV. self-regulation.

A: I, II and III.
B: I, II and IV.
C: I, III and IV.
D: II, III and IV.
E: I, II, III and IV.

Learning Response C: Correct. The regulation of these genes involves transcriptional and post-transcriptional effects. They are controlled by their own gene product and by the product of genes adjacent on the chromosome. They are also alternatively spliced.

---

373. SUBJECT AREA: Cellular Mechanisms of Development
QUESTION: Connective tissue can control the following aspects of the development of migrating cells EXCEPT:

A: the paths along which they migrate.
B: the sites at which they settle.
C: their ability to proliferate.
D: their survival.

E: all of the above.

Learning Response E: Correct. A migratory cell's path, settlement, proliferation, differentiation, and survival can all be controlled by the connective tissue that it colonizes.

---

374. SUBJECT AREA: Cellular Mechanisms of Development
QUESTION: All of the following are true of the role of cadherins in development, EXCEPT:

A: changes in their expression correlate with changing patterns of cell association.
B: they are involved in cell-cell adhesion.
C: they control the formation and dissolution of epithelial sheets.
D: they have been implicated in imaginal disc folding.
E: they translate position information into morphogenic movement.

Learning Response D: Correct. Due to correlations between the expression of cadherins, which mediate cell-cell adhesion, and the changing patterns of cell association seen during development, they are thought to translate information about position into cellular movement. Specifically, they are thought to be involved in the formation and dissolution of epithelial sheets. Integrins, which mediate cell-matrix association, have been implicated in imaginal disc folding.

---

375. SUBJECT AREA: Cellular Mechanisms of Development
QUESTION: The following are true of the connective tissue in vertebral limbs EXCEPT:

A: it derives from the somites of the early embryo.
B: it develops from the mesenchyme.
C: it forms the bone and tendons.
D: it is composed of fibroblasts and related cells.
E: it serves as host to immigrant populations of cells of other origin.

Learning Response A: Correct. The connective tissue of the limb derives from the mesenchyme, which comes from the lateral plate mesoderm flanking the somites. Composed of fibroblasts and related cells, it forms bone, cartilage, tendons, ligaments, the dermis, and other tissues, but other tissues of the limb, particularly the muscle, derive from cells of non-mesenchymal origin that migrate through the connective tissue to take up residence and form tissue.

---

376. SUBJECT AREA: Differentiated Cells and the Maintenance of Tissues
QUESTION: Of the cells usually necessary for a tissue to be maintained, all of the following are present in the tissue early in development, EXCEPT:

A: endothelial.
B: fibroblasts.
C: macrophages.
D: melanocytes.
E: nerve cells.

Learning Response C: Correct. While nerve and endothelial cells as well as melanocytes invade developing tissues early in development, macrophages and lymphocytes are continually brought to the tissue through the circulation.

---

377. SUBJECT AREA: Differentiated Cells and the Maintenance of Tissues
QUESTION: Which of the following is true of differentiated cells?

A: their characteristics are lost when cells are cultured.
B: their development is independent of environment.
C: their final character is irreversible.
D: their interconversions are narrowly restricted.
E: they are devoid of cell memory.

Learning Response D: Correct. The differentiated state of cells represents the development of cell memory influenced by environmental effects. It can be maintained in culture after the cells have been removed from their environment and while modulations in the final character are possible, they are narrowly restricted.

378. Subject Area: Differentiated Cells and the Maintenance of Tissues
Question: Which of the following is true of all cells in differentiated tissue?

A: they are generated in the embryo and retained throughout life.
B: they are of a permanent nature.
C: they are replaced if lost.
D: they are subject to turnover.
E: none of the above are true of all cells.

Learning Response E: Correct. While in most tissues there is a continual cycle of cell death and replacement, in some tissues cells that generate in the embryo are maintained, and if lost can not be regenerated.

---

379. Subject Area: Differentiated Cells and the Maintenance of Tissues
Question: Which of the following are true of most permanent cells?

I. they are of constant size and structure.
II. they are not replaced.
III. they cease protein synthesis.
IV. they do not divide.

A: I and II.
B: II and IV.
C: I, II and III.
D: I, II and IV.
E: I, II, III, and IV.

Learning Response B: Correct. While most permanent cells neither divide nor can be replaced if lost, they continue synthesis of protein and can be observed to change shape and size.

---

380. Subject Area: Differentiated Cells and the Maintenance of Tissues
Question: All the following are true of replacement of cell populations, EXCEPT:

A: it can be stimulated by tissue damage.
B: it can involve a change of phenotype.
C: it can involve stem cells.
D: it can occur by duplication of existing cells.
E: it occurs at similar rates among tissues.

Learning Response E: Correct. Renewal of cells involves either duplication of existing cells or differentiation of dividing stem cells, with the corresponding change in phenotype. This occurs at different rates in different tissues and can be stimulated by the need of the tissue.

---

381. SUBJECT AREA: Differentiated Cells and the Maintenance of Tissues
QUESTION: The following are true of liver cell renewal, EXCEPT:

A: it is signaled by a soluble mediator.
B: it is stimulated by cell loss.
C: it occurs at a rapid but controlled rate in normal liver.
D: it occurs by a system analogous to that in the kidney.
E: it occurs through duplication of existing cells.

Learning Response C: Correct. In normal tissue, the rate of liver regeneration is slow and controlled. Mediated by duplication, it can be stimulated by the removal of a portion of the liver. This even allows growth of an undamaged liver connected through the circulation to the damaged liver.

---

382. SUBJECT AREA: Differentiated Cells and the Maintenance of Tissues
QUESTION: All of the following are true of the regeneration of most tissues, EXCEPT:

A: it involves the reproduction of numerous cell types.
B: it is a simple process of cell duplication.
C: it is best studied in cases of tissue injury.
D: it must be highly coordinated.
E: its improper progression can cause permanent damage.

Learning Response B: Correct. Tissue regeneration involves the coordinate growth of many different cell types. When improperly coordinated as in the case of certain tissue injuries, it can result in permanent damage to the tissue.

---

383. SUBJECT AREA: Differentiated Cells and the Maintenance of Tissues
QUESTION: The following are true of the proliferation of blood vessel endothelial cells, EXCEPT:

A: continually generates new vessels.
B: involves a short cell lifespan.
C: is responsive to signals from surrounding tissue.
D: occurs by duplication.
E: replaces damaged neighboring cells.

Learning Response B: Correct. The duplication of these endothelial cells, which normally have a lifespan of months or years, causes the replacement of damaged neighboring cells. Also, in response to signals from tissues without sufficient vasculature, it will lead to generation of new vessels.

---

384. SUBJECT AREA: Differentiated Cells and the Maintenance of Tissues
QUESTION: The following are true of cell renewal by stem cells, EXCEPT:

A: it involves a choice between two developmental courses.
B: it involves a dedifferentiation step.
C: it involves a different geometry than simple division.
D: it requires a population of undifferentiated cells.
E: stem cells must have unlimited division potential.

Learning Response B: Correct. In this process undifferentiated stem cells divide to produce daughters which will either remain as stem cells or terminally differentiate. The stem cells must then continue dividing over the life span of the animal. This also produces a directionality to cell generation that is not present in the case of simple division.

385. SUBJECT AREA: Differentiated Cells and the Maintenance of Tissues
QUESTION: Which of the following refers to stem cells that give rise to a small number of cell types?

A: multipotent.
B: oligopotent.
C: pluripotent.
D: unipotent.
E: none of the above.

Learning Response B: Correct. Stem cells are referred to as unipotent if they produce a single cell type, oligopotent if they give rise to a small number of cell types, and pluripotent if they give rise to many cell types.

386. SUBJECT AREA: Differentiated Cells and the Maintenance of Tissues
QUESTION: All the following are characteristics of squames, EXCEPT:

A: they are keratinocytes.
B: they contain involucrin.
C: they divide slowly.
D: they form a regular geometric arrangement.
E: they have lost their organelles.

Learning Response C: Correct. The keratinocytes which make up the epidermis gradually progress to become squames. They develop a tough cross linked layer of involucrin, lose their organelles and die. These dead cells are found to form a regular geometric pattern.

387. SUBJECT AREA: Differentiated Cells and the Maintenance of Tissues
QUESTION: Immortal epidermal stem cells:

A: always divide asymmetrically.
B: cannot repopulate a damaged area.
C: give rise to daughter cells that will differentiate.
D: have been transformed by an oncogene.

E: never increase in number.

Learning Response C: Correct. Immortal stem cells represents a line of cells that give rise to both stem cells and cells which will differentiate. That they always divide asymmetrically and hence never increase in number is disproven by their ability to repopulate damaged regions of the epidermis.

---

388. SUBJECT AREA: Differentiated Cells and the Maintenance of Tissues
QUESTION: Which of the following types of blood cells have a multilobed nucleus?

A: basophils.
B: eosinophils.
C: erythrocytes.
D: macrophages.
E: neutrophils.

Learning Response E: Correct. Neutrophils are the most common type of granulocyte, and are sometimes called polymorphonuclear leucocytes because of their unusual nuclear morphology.

---

389. SUBJECT AREA: Differentiated Cells and the Maintenance of Tissues
QUESTION: Hemopoiesis involves all the following EXCEPT:

A: bone marrow.
B: complex regulation.
C: pluripotent stem cells.
D: signaling from the periphery.
E: the inflammatory response.

Learning Response E: Correct. While the inflammatory response produces signals which affect the hemopoietic process, that process does not directly involve the inflammatory response. Pluripotent stem cells in the bone marrow differentiate under complex regulatory control to produce the cells of the blood.

---

390. SUBJECT AREA: Differentiated Cells and the Maintenance of Tissues
QUESTION: Colony-forming cells:

A: are diverse in nature.
B: can survive x-irradiation.
C: descend from differentiated blood cells.
D: develop in the bone marrow.
E: are blood cells grown in culture.

Learning Response A: Correct. When a mouse is irradiated to destroy its own hemopoietic cells and injected with cells from a donor, stem cells deposit in the spleen and develop into colony forming cells. The nature of these cells depend on the potency and lineage of the stem cell giving rise to the colony.

---

391. SUBJECT AREA: Differentiated Cells and the Maintenance of Tissues
QUESTION: All of the following are true of committed progenitor cells, EXCEPT:

A: they are committed to a single developmental fate.
B: they develop from pluripotent stem cells.
C: they develop into terminally differentiated cells.
D: they divide a limited number of times.
E: they divide rapidly.

Learning Response A: Correct. Pluripotent stem cells give rise to cells which are irreversibly committed to either lymphoid or myeloid development. These cells undergo a limited number of rapid divisions before developing into terminally differentiated cells.

---

392. SUBJECT AREA: Differentiated Cells and the Maintenance of Tissues
QUESTION: The following are true of the development of erythrocytes EXCEPT:

A: it can be studied in culture.
B: it involves erythropoietin-dependent burst-forming cells.
C: it involves signaling from the kidney.

D: it requires production of hemoglobin.
E: it requires several days.

Learning Response B: Correct. Erythropoietin produced by the kidney will stimulate the production of red blood cells. This process, which can be studied in culture, involves the development of burst-forming cells which are committed to the erythroid lineage. These give rise to colony forming cells which are erythropoietin dependent and in turn develop into hemoglobin-contained erythrocytes.

---

393. SUBJECT AREA: Differentiated Cells and the Maintenance of Tissues
QUESTION: Which type of muscle cell derives from the ectoderm?

A: heart muscle cells.
B: myoepithelial cells.
C: skeletal muscle cells.
D: smooth muscle cells.
E: all muscle cells are of mesodermal origin.

Learning Response B: Correct. While most muscle cells are of mesodermal origin, the myoepithelial cells lie in epithelia and are of ectodermal origin.

---

394. SUBJECT AREA: Differentiated Cells and the Maintenance of Tissues
QUESTION: Which of the following is NOT true of skeletal muscle subtypes?

A: they are found in distinct muscles.
B: they can change character under inappropriate influences.
C: they express different amounts of myoglobin.
D: they express different copies of variant genes.
E: they have different oxygen requirements.

Learning Response A: Correct. Different muscle types have different amounts of myoglobin reflecting different oxygen requirements, but can be found side by side in the same muscle. By altering their nerve stimulation, they can be caused to switch characteristics by altering expression of their genes.

---

395. SUBJECT AREA: Differentiated Cells and the Maintenance of Tissues
QUESTION: Which of the following is not a member of the connective-tissue cell family?

A: bone cells
B: cartilage cells.
C: fat cells.
D: fibroblasts.
E: all the above are members of this family.

Learning Response E: Correct. While this family includes fibroblasts, cartilage cells and bone cells which each secretes collagenous extracellular matrix material, it also includes fat cells and smooth muscle cells.

---

396. SUBJECT AREA: Differentiated Cells and the Maintenance of Tissues
QUESTION: The following are true of adipocytes, EXCEPT:

A: they can be studied in culture.
B: their differentiation involves production of specific enzymes.
C: their differentiation involves soluble factors.
D: their differentiation is affected by cell shape.
E: their differentiation is irreversible.

Learning Response E: Correct. Fat cell differentiation, which can be studied in culture, requires stimulation by certain soluble factors but can also be affected by the shape of the cell. This differentiation which is partially reversible involves the activation of specific enzymes, the production of lipid droplets and their fusion into a single lipid mass.

---

397. SUBJECT AREA: Differentiated Cells and the Maintenance of Tissues
QUESTION: Which of the following is true of bone?

A: it contains irregularly layered fibrils of collagen.
B: it is degraded by apposition.
C: it is full of channels and cavities.
D: it is immutable once it is formed.

E: it is primarily composed of calcium phosphate.

Learning Response C: Correct. Bone is formed by approximately equal amounts of calcium phosphate and regular layers of collagen. It is continually remodeled by the cells which reside within its many channels and cavities. These both destroy old bone and deposit new bone through apposition.

---

398. SUBJECT AREA: Differentiated Cells and the Maintenance of Tissues
QUESTION: Which of the following describes osteoclasts?

A: they develop from monocytes.
B: they develop into bone cells.
C: they form gap junctions.
D: they perform apposition.
E: they secrete cartilage.

Learning Response A: Correct. Osteoclasts derive from the same lineage as macrophages. They arrive in bone and fuse into multinucleated cells. They then begin to digest the bone which is replaced by osteocytes following in their wake.

---

399. SUBJECT AREA: Differentiated Cells and the Maintenance of Tissues
QUESTION: All the following are true of the classification of cells in the human body, EXCEPT:

A: the fineness of their subdivisions is somewhat arbitrary.
B: their classification represent a continuum of characteristics.
C: their traditional classification is based on shape and structure.
D: there are as many as 200 different types.
E: there is some variation within cell types.

Learning Response B: Correct. Cells were traditionally classified by shape and structure. This classification results in some 200 cell types, but the fineness of subdivisions is somewhat arbitrary, and there remains some variation within cell types. This system does not, however, represent a continuum of characteristics for each cell type has distinctive traits.

---

400. SUBJECT AREA: The Immune System
QUESTION: Which of the following mature in the thymus?

I. B cells.
II. red blood cells.
III. T cells.

A: I.
B: III.
C: I and III.
D: II and III.
E: I, II and III.

Learning Response B: Correct. T cells are produced and mature in the thymus, while B cells and erythrocytes are produced by the bone marrow in mammals.

---

401. SUBJECT AREA: The Immune System
QUESTION: Which of the following is inconsistent with the instructional hypothesis of antibody formation?

A: antibodies are proteins.
B: antibodies are multivalent.
C: antibodies can activate complement.
D: denatured antibody can often refold into its original binding site.
E: there are several different classes of antibodies.

Learning Response D: Correct. Antibodies are multivalent proteins of five classes and can activate complement, but these would be consistent with either explanation of antibody formation. However, the ability of antibodies that have been denatured to refold into their original structures without the presence of antigen is inconsistent with instructional theories.

---

402. SUBJECT AREA: The Immune System
QUESTION: All of the following are secondary lymphoid organs, EXCEPT:

A: adenoid.
B: appendix.
C: spleen.
D: thymus.
E: tonsil.

Learning Response D: Correct. After being produced in the primary lymphoid organs, the thymus and bone marrow, lymphocytes migrate to the secondary lymphoid organs.

---

403. SUBJECT AREA: The Immune System
QUESTION: All of the following are true of tolerance, EXCEPT:

A: it can be induced in adults.
B: it involves elimination of lymphocytes.
C: it is inborn.
D: it requires continual presence of the antigen.
E: its breakdown can cause autoimmune disease.

Learning Response C: Correct. The development of tolerance, while inducible in adults, occurs as the immune system after birth. By elimination of lymphocytes which respond to self antigens, tolerance to these antigens is developed. This tolerance requires the continual presence of the antigen, and its breakdown is thought to lead to such diseases as *Myasthenia gravis* and arthritis.

---

404. SUBJECT AREA: The Immune System
QUESTION: Which class of Ig has the highest valence?

A: IgA.
B: IgD.
C: IgE.
D: IgG.

E: IgM.

Learning Response E: Correct. IgM, with its 5 four-chain subunits, for 10 binding sites, has the greatest valence, the others having only 2-4 binding sites.

---

405. SUBJECT AREA: The Immune System
QUESTION: Which class of Ig has four subclasses?

A: IgA.
B: IgD.
C: IgE.
D: IgG.
E: IgM.

Learning Response D: Correct. Having four different heavy chains, IgG is divided into four different subclasses on this basis. Each of the others has only a single type of heavy chain.

---

406. SUBJECT AREA: The Immune System
QUESTION: Which Ig class is expressed first in cells of the B lineage?

A: IgA.
B: IgD.
C: IgE.
D: IgG.
E: IgM.

Learning Response E: Correct. In addition to being the major class of secreted Ig, IgM is the first class produced by developing B cells, which only later begin to produce IgD.

---

407. SUBJECT AREA: The Immune System
QUESTION: The biological or effector functions of antibodies, such

as complement activation or sensitization of mast cells for anaphylactic reactions, are directly due to which portion of the antibody molecule?

A: Fab.
B: $F(ab')_2$.
C: Fc.
D: L Chain.
E: V chain.

Learning Response C: Correct. While not binding antigen directly (the role of the Fab region) the Fc portion of an antibody is involved in binding interactions with other components of the immune system, resulting in complement, macrophage, neutrophil, and mast cell activation.

---

408. SUBJECT AREA: The Immune System
QUESTION: Which Ig class is transported across mucosal barriers?

A: IgA.
B: IgD.
C: IgE.
D: IgG.
E: IgM.

Learning Response A: Correct. IgA binds to a secretory component of cells of epithelial mucosa lining the intestine, bronchi, and other tissues, is transported to the lumenal face of these cells, and released.

---

409. SUBJECT AREA: The Immune System
QUESTION: Which class of Ig is associated with allergic reactions?

A: IgA.
B: IgD.
C: IgE.
D: IgG.
E: IgM.

Learning Response C: Correct. The binding of antigen to cell-bound IgE results in the release of amines, which in turn result in the symptoms of an allergic reaction.

---

410. SUBJECT AREA: The Immune System
QUESTION: Antibody-antigen interactions involve formation of all of the following EXCEPT:

A: carbon-carbon bonds.
B: electrostatic bonds.
C: hydrogen bonds.
D: hydrophobic interactions.
E: van der Waals interactions.

Learning Response A: Correct. The interaction of antibodies and antigens involves a variety of non-covalent interactions, but no covalent bonds.

---

411. SUBJECT AREA: The Immune System
QUESTION: Which of the following heavy chains can pair with λ light chains?

A: μ.
B: γ.
C: α.
D: ε.
E: all of the above.

Learning Response E: Correct. Both the λ and κ light chains can pair with any of the heavy chains. However, within a single antibody tetramer, only a single type of heavy chain and a single type of light chain will be used.

---

412. SUBJECT AREA: The Immune System
QUESTION: A typical Ig domain:

A: binds to an Fc receptor.
B: has a variable and a constant domain.
C: is composed of α helixes.

D: is composed of two layers of β-pleated sheets.
E: is rich in proline

Learning Response D: Correct. A single Ig domain, comprising either a single variable or a single constant domain, is formed by the juxtaposing of two layers of β sheets, cross-linked to each other by disulfide bonding.

---

413. SUBJECT AREA: The Immune System
QUESTION: Which of the following regions is not part of a light chain?

A: C.
B: D.
C: J.
D: V.
E: all of the above are part of light chains.

Learning Response B: Correct. While the heavy chains of antibodies are produced by joining the constant (C) region to one of several variable (V), junction (J), and diversity (D) domains, only V and J domains are joined to the light chain C region.

---

414. SUBJECT AREA: The Immune System
QUESTION: Diversity of the possible antibodies produced by a cell is increased by each of the following, EXCEPT:

A: combination of differentially spliced light and heavy chains.
B: imprecise splicing of gene regions.
C: mixing of paternal and maternal light chains.
D: selection of different V-J-D regions.
E: use of alternative light chains.

Learning Response C: Correct. While a cell may mix a paternal heavy chain with maternal light chain, or vice versa, both light chains (and both heavy chains) will come from the same chromosome, and not one from each.

415. SUBJECT AREA: The Immune System
QUESTION: Switching of antibody production by B cells involves:

I. use of different chromosomal copies.
II. use of different classes of heavy chains.
III. use of different V-J-D domains.
IV. use of secreted rather than membrane form of heavy chain.

A: I and II.
B: I and III.
C: II and III.
D: II and IV.
E: III and IV.

Learning Response D: Correct. The B cell switch involves the use of other constant regions, either in terms of class or within a class from membrane bound to secreted forms. These changes can involve either a change in RNA splicing, or in the DNA itself.

---

416. SUBJECT AREA: The Immune System
QUESTION: Activation of the complement response involves all of the following, EXCEPT:

A: alternative pathways of activation.
B: assembly of a large attack complex.
C: cleavage of C3.
D: recognition of microbial infection.
E: stimulation of antibody production.

Learning Response E: Correct. The complement response involves microbial activation of alternative pathways of soluble proteins, resulting in the cleavage of C3. This, in turn, promotes the assembly of the late complement components into an attack complex, which causes microbe lysis.

---

417. SUBJECT AREA: The Immune System

QUESTION: Of the following effects of the cleavage of C3, all are mediated by C3b, EXCEPT:

A: activation of C5
B: activation of phagocytosis in macrophages.
C: secretion of histamine.
D: stimulation of the alternative pathway.
E: all of the above are C3b mediated.

Learning Response C: Correct. While C3b causes the activation of its own production, of C5, and of macrophages and neutrophils, C3a is responsible for smooth muscle contraction and histamine production seen with C3 cleavage.

---

418. SUBJECT AREA: The Immune System
QUESTION: Immunoglobulins and T cell receptors share all of the following features EXCEPT:

A: they are anchored by disulfide bonds.
B: they are composed of multiple chains.
C: they bind soluble antigen.
D: they have constant and variable regions.
E: they use V-D-J recombination to generate diversity.

Learning Response C: Correct. While T cell receptors consist of antibody-like heterodimers with similar V-D-J recombination, and with variable and constant chains folded by disulfide bridges, they can only bind antigen when it is bound or attached to another cell, and not when soluble.

---

419. SUBJECT AREA: The Immune System
QUESTION: Which of the following T cell surface molecules is involved in mediating cell-to-cell adhesion?

A: CD1.
B: CD2.
C: CD3.
D: CD5.

E: CD25.

Learning Response B: Correct. By binding to LFA-3 on target cells, CD2 promotes adhesion between the T cell and its target.

---

420. SUBJECT AREA: The Immune System
QUESTION: The action of cytotoxic T cells can involve all of the following, EXCEPT:

A: binding of antibody.
B: cytoskeletal rearrangements in the T cell.
C: killing of host cells.
D: production of perforins.
E: recognition of viral antigens.

Learning Response A: Correct. Lysis of virally infected cells involves the recognition of viral antigens on the host cell by the T cell receptor. This causes receptor aggregation and a redirection of the T cell's production apparatus toward the infected cell. By the use of perforins, as well as by perforin independent methods, the host cell is then lysed.

---

421. SUBJECT AREA: The Immune System
QUESTION: All of the following are true of the MHC molecule classes, EXCEPT:

A: their genes are highly polymorphic.
B: they are dimers.
C: they are encoded by single gene loci.
D: they are involved in transplant rejection.
E: they focus T cells on virally infected cells.

Learning Response C: Correct. MHC molecules, which are involved in transplant rejection, are heterodimers which present viral antigen to T cells. They are encoded by at least seven highly polymorphic genes.

422. SUBJECT AREA: The Immune System
QUESTION: Which of the following are recognized by cytotoxic T cells?

A: antigen presented by antigen presenting cells.
B: foreign class II MHC glycoproteins.
C: infected cells.
D: native viral proteins.
E: soluble fragments of viral proteins.

Learning Response C: Correct. Cytotoxic T cells recognize the fragments of viral proteins that are presented by the class I MHC molecules of infected cells. They also bind to class I MHC molecules on foreign (transplant) cells.

---

423. SUBJECT AREA: The Immune System
QUESTION: All of the following are true of helper T cells, EXCEPT:

A: they all secrete interleukins and y-interferon.
B: they are killed by HIV.
C: they are required for a B cell response to most antigens.
D: they are stimulated by antigen presenting cells.
E: they interact with class II MHC glycoproteins.

Learning Response A: Correct. Helper T cells are stimulated by viral antigen presented by the class II MHC glycoproteins of antigen presenting cells. This induces different classes to release different combinations on interleukins, and in some cases y-interferon, to promote the response of B cells. It is this class of cells which is infected by HIV viruses.

---

424. SUBJECT AREA: The Nervous System
QUESTION: All of the following are true of a synapse, EXCEPT:

A: it conveys signals from an axon.
B: it involves a chemical neurotransmitter.

C: it involves a significant turnover of signaling molecules.
D: it is more adaptable than a gap junction.
E: it is the sole mechanism of conducting electrical signals between cells.

Learning Response E: Correct. A synapse provides a more versatile mechanism for transmitting an electrical signal from an axon to the cell it is signaling than does a gap junction, which is also used for electrical coupling. It involves secretion of a soluble messenger that must be degraded rapidly after release, causing a high amount of biochemical activity.

---

425. SUBJECT AREA: The Nervous System
QUESTION: All of the following are distinctions between fast axonal transport and fast retrograde transport, EXCEPT:

A: fast axonal transport carries larger vesicles.
B: fast axonal transport often carries synaptic vesicles.
C: fast retrograde transport is slower.
D: they are driven by different motors.
E: they are oriented in opposite directions.

Learning Response A: Correct. These two intercellular transport mechanisms carry out opposing motion in the neuron. Using a different motor, fast axonal transport is faster, but carries smaller vesicles, including synaptic vesicles, away from the cell body, while fast retrograde transport proceeds toward the cell body.

---

426. SUBJECT AREA: The Nervous System
QUESTION: The following is true of all gated channels, EXCEPT:

A: their permeability is regulated.
B: they are ion selective.
C: they are opened by changes in voltage.
D: they are transmembrane proteins.
E: they control the membrane potential.

Learning Response C: Correct. These transmembrane channels control the membrane potential through their regulated opening and closing. Selective for a single ion or group of ions, their opening is regulated by either voltage or by specific ligands.

---

427. SUBJECT AREA: The Nervous System
QUESTION: Which of the following describes the role of sodium and potassium in action potentials?

A: a ligand-gated channel is opened by sodium, and then closed by potassium.
B: the potassium channels respond with more rapid kinetics than the sodium channels.
C: the potassium overwhelms the sodium influx.
D: the sodium channels are inactivated by potassium.
E: they spontaneously initiate an action potential.

Learning Response C: Correct. A momentary depolarization of the cell membrane initiates the action potential by opening the voltage-gated sodium channel, which eventually closes spontaneously. Voltage-gated potassium channels, which respond with slower kinetics, open to cause a potassium efflux which overwhelms the sodium influx, and reestablishes the membrane polarization.

---

428. SUBJECT AREA: The Nervous System
QUESTION: The following are true of oligodendrocytes, EXCEPT:

A: they allow faster movement of action potentials.
B: they conserve neuron metabolic energy.
C: they form a continuous insulation.
D: they prevent most leakage from neurons.
E: they wrap around multiple neurons.

Learning Response C: Correct. Unlike Schwann cells, oligodendrocytes wrap around several neurons. They provide insulation that prevents leakage and allows a faster movement of signal. This insulation in not contiguous, however, leaving small gaps where the machinery for action potentials is concentrated, saving the neuron the energy of covering the entire membrane with the necessary proteins.

---

429. SUBJECT AREA: The Nervous System
QUESTION: Which of the following is true of ligand-gated ion channels?

A: they are specialized gap junctions.
B: they bind directly to neurotransmitters.
C: they conduct action potentials along a cell.
D: they involve a second messenger.
E: they respond slowly to activation.

Learning Response B: Correct. Ligand-gated channels are involved in transmitting electrical signals across a synapse. In response to the binding of neurotransmitters, they open rapidly, resulting in a flow of ions into the postsynaptic cell.

---

430. SUBJECT AREA: The Nervous System
QUESTION: All of the following are true of the axonal calcium channels at the neuromuscular junction, EXCEPT:

A: their action can be mimicked through microinjection.
B: they allow a greater ion flux than sodium or potassium channels.
C: they are voltage-gated.
D: they open in response to the propagating action potential.
E: they stimulate the release of neurotransmitter.

Learning Response B: Correct. When the action potential driven by sodium and potassium reaches the axon terminal of the presynaptic cell, it triggers the opening of voltage-gated calcium channels. While the influx of calcium is no greater than that seen through other channels, it is more dramatic due to significantly lower calcium levels in the cell. As proven by microinjection experiments, it is this influx that leads to release of neurotransmitter into the synapse by the axon.

---

431. SUBJECT AREA: The Nervous System

QUESTION: A miniature synaptic potential:

A: activates the presynaptic cell.
B: initiates an action potential.
C: occurs during calcium channel activation.
D: represents the effect of one secretory vesicle.
E: results from a single molecule of acetylcholine.

Learning Response D: Correct. Axonal secretory vesicles are subject to occasional spontaneous fusion events with the plasma membrane. This causes the release of their content, about 5000 acetylcholine molecules, and a brief, membrane depolarization of insufficient magnitude to cause an action potential in the postsynaptic membrane. The calcium influx into the presynaptic cell results in the simultaneous release of about 300 such vesicles.

---

432. SUBJECT AREA: The Nervous System
QUESTION: Removal of acetylcholine in the synaptic cleft occurs through which of the following?

I. destruction by the gated channel.
II. diffusion from the cleft.
III. hydrolysis in the cleft.

A: II.
B: III.
C: I and II.
D: II and III.
E: I, II and III.

Learning Response D: Correct. The spike of acetylcholine released at a synapse is eliminated through its direct destruction by acetylcholine esterase, and by its rapid diffusion.

---

433. SUBJECT AREA: The Nervous System
QUESTION: The following are true of postsynaptic potentials, EXCEPT:

A: their magnitude falls off with distance from the site of the synapse.
B: they are subject to spatial summation at the cell body.
C: they are subject to temporal summation at the cell body.
D: they can be either excitatory or inhibitory.
E: they generally do not trigger an action potential, when coming alone.

Learning Response C: Correct. Postsynaptic potentials can be either excitatory or inhibitory. A PSP is usually not sufficient to cause an action potential. Unless local temporal summation or spatial summation at the cell body allow a certain threshold to be reached, the local effect will fall away with distance.

---

434. SUBJECT AREA: The Nervous System
QUESTION: The process of adaptation:

A: follows calcium channel activation.
B: is mediated by a ligand-gated channel.
C: is mediated by sodium.
D: makes a cell easier to stimulate.
E: makes membrane depolarization easier.

Learning Response A: Correct. The opening of voltage-gated calcium channels causes activation of a calcium-gated potassium channel. This increases the permeability of the membrane to potassium, making the membrane harder to depolarize, and thus makes it harder to stimulate.

---

435. SUBJECT AREA: The Nervous System
QUESTION: The following are true of non-channel-linked receptors EXCEPT:

A: their action is slower than channel receptors.
B: their actions in a cell tend to be spatially diffuse.
C: they can act through several different mechanisms.
D: they can involve G proteins.

E: all of the above are true.

Learning Response E: Correct. Non-channel receptors can act by many mechanisms. They either have their own catalytic activity, which can directly modify channels, or they activate G proteins. These, in turn, can directly activate channels or generate second messengers which can act directly or through kinases to mediate their effects on channels. This causes both slower and more diffuse action than the direct opening of channels by ligands.

---

436. SUBJECT AREA: The Nervous System
QUESTION: Which of the following is true of neuropeptides?

A: their action is limited to the synapse at which they are released.
B: they are contained in smaller vesicles than other neurotransmitters.
C: they are long polypeptide chains.
D: they are produced by cleavage of a precursor.
E: they are produced in axons.

Learning Response D: Correct. Neuropeptides are small peptides produced in the cell body and cleaved from a larger protein chain. Transported in large vesicles to the axons, they are able to diffuse allowing for action over a broader area.

---

437. SUBJECT AREA: The Nervous System
QUESTION: Which of the following is true of long term potentiation?

A: it depends only on actions at a single synapse.
B: it results from single postsynaptic potentials.
C: it lasts for a few days.
D: it involves the local action of calcium.
E: it prevents additional activation of the synapse.

Learning Response D: Correct. Repetitive bursts of stimulation cause long term potentiation at any synapse which is stimulated during the resulting strong depolarization. This sensitizes the synapses to further stimulation through a calcium dependent mechanism, which effect can last for months.

---

438. SUBJECT AREA: The Nervous System
QUESTION: All of the following are shared by all sensory cells, EXCEPT:

A: they all serve as transducers.
B: they are neurons.
C: they establish receptor potentials.
D: they produce graded responses.
E: they use different types of receptors.

Learning Response B: Correct. Sensory cells use various receptor types to establish a graded response, in the form of a receptor potential, thus transducing the received stimulus into an electrical signal. They then either pass this signal directly to the central nervous system (if the cells themselves are neurons), or pass the signal across a synapse to a neuron that will then transmit the signal on.

---

439. SUBJECT AREA: The Nervous System
QUESTION: Photoreception in the mammalian eye involves all of the following, EXCEPT:

A: action through interneurons.
B: an inhibitory neurotransmitter.
C: depolarization by light.
D: reduction of cGMP.
E: stimulation of G proteins.

Learning Response C: Correct. Activation of rhodopsin by light leads to the activation of transducin, a G protein. This stimulates a cGMP phosphodiesterase, and the resulting loss of stimulation to a cGMP-stimulated sodium channel leads to a hyperpolarization of the membrane. This hyperpolarization results in a reduction in the release of an inhibitory neurotransmitter, and an increase in interneuron signaling.

440. SUBJECT AREA: The Nervous System

QUESTION: All of the following are characteristics of the neural tube, EXCEPT:

A: it gives rise to neurons and glial cells.
B: it initially consists of a single cell layer.
C: it is epithelial in origin.
D: its cells can undergo infinite cell divisions.
E: its regions each have their own program of cell divisions.

Learning Response D: Correct. The neural tube is a developmental structure that originally consists of a single layer of epithelial cells, and gives rise to the glial cells and neurons of the central nervous system. These are produced from the different regions of the neural tube, which each have their own finite program of cell divisions.

441. SUBJECT AREA: The Nervous System

QUESTION: Changing the final location of cells, as in the *reeler* mutation, appears to significantly alter which of the following:

I. the connections that they make with other neurons.
II. the morphology which they develop.
III. their birthday.
IV. their migration to their final location.

A: IV.
B: I and IV.
C: II and IV.
D: III and IV.
E: I, II, III and IV.

Learning Response A: Correct. While changing the final location of the cells will obviously effect their migration, it does not effect birthday, morphology, or most of the connections that the cells would normally form.

---

442. Subject Area: The Nervous System
Question: All of the following are true of a neuron growth cone, EXCEPT:

A: it has many filopodia.
B: it is sensitive to drugs that interfere with microtubules.
C: it is the mechanism for axon growth.
D: its course can be directed by contact guidance.
E: the glycoproteins which guide its growth remain unidentified.

Learning Response E: Correct. Through the extension of filopodia by a mechanism involving microtubule polymerization, the growth cone extends axons and dendrites. This growth is guided by specific glycoproteins in the extracellular matrix, including N-CAM and L1 glycoprotein.

---

443. Subject Area: The Nervous System
Question: All of the following regulate the pattern of innervation seen in the mature organism, EXCEPT:

A: cell death in the absence of target tissue.
B: remodeling of contacts over time.
C: secretion of factors by their target tissues.
D: specificity of nerves for their targets.
E: all of the above are involved.

Learning Response E: Correct. The pattern seen in adults is a result of the original innervation as well as remodeling. The nerves appear to have an initial specificity for a target tissue, and are directed toward the appropriate tissue by neurotropic factors. Those ganglia which fail to innervate a target tissue will tend to regress.

---

444. Subject Area: The Nervous System
Question: The following accompany the formation of a neuromuscular synapse, EXCEPT?

A: a specialized region of basal lamina.
B: formation of structural and biochemical specializations.
C: migration of acetylcholine receptors towards the synapse site.
D: modifications of the muscle prior to formation of the synapse.
E: rapid maturation of initial contacts.

Learning Response E: Correct. This process involves signaling between the neuron and muscle before contact is made. The junction form special regions in the basal lamina, and initiates the process of maturation, which involves numerous changes to both neuron and muscle cells, including the migration of receptors to the contact site, and takes several days.

---

445. SUBJECT AREA: The Nervous System
QUESTION: Denervation of a muscle will induce the following changes, EXCEPT:

A: concentration of receptor at the synapse.
B: loss of electrical activity.
C: production of sprouting factor.
D: receptivity to new enervation.
E: synthesis of new receptors.

Learning Response A: Correct. The loss of innervation to a muscle results in a loss of electrical stimulation to the muscle. This in turn leads to several changes. The muscle produces more receptors, and these are no longer concentrated at synapses. They also produce sprouting factors, which advertise their receptivity to new innervation.

---

446. SUBJECT AREA: The Nervous System
QUESTION: Which of the following describes an aspect of the process of competitive elimination of synapses?

A: competition of only synapses in close proximity.
B: death of surplus embryonic motor neurons.
C: elimination of branches which have recently fired.

D: elimination of all but one synapse from each neuron.
E: random elimination of axon branches.

Learning Response A: Correct. The process of competitive elimination of synapses involves the elimination of all but a single synapse from a muscle cell. It would appear that the firing at a synapse will result in the release of a factor which leads to the rejection of adjacent synapses which have not recently fired, eventually eliminating all but one.

---

447. SUBJECT AREA: The Nervous System
QUESTION: Which of the following describes the connections of the mammalian visual system?

A: each eye is enervated by one side of the brain.
B: the pattern is fluid throughout life, depending on input.
C: the pattern is mature at birth.
D: they involve separate nerves for each eye.
E: they result in a banded pattern of innervation in the visual cortex.

Learning Response E: Correct. The innervation of the mammalian eye experiences a sensitive period during childhood, before setting into a rigid pattern. The inputs from the two eyes establish a pattern of alternate dominance columns on the visual cortex, as well as innervating a certain number of neurons responsible for stereo vision from both eyes.

---

448. SUBJECT AREA: Special Features of Plant Cells
QUESTION: All of the following are true of nitrogen fixation, EXCEPT:

A: it involves the action of nitrogenase.
B: it is performed by soil fungi.
C: it requires a great deal of energy.
D: it requires the coordinated expression of many genes.
E: legumes often form symbiotic associations with

Rhizobium.

Learning Response B: Correct. Nitrogen fixation requires a great deal of energy and is performed for some plants by the bacterium, *Rhizobium*. This process requires many genes, but is performed by the enzyme, nitrogenase.

---

449. SUBJECT AREA: Special Features of Plant Cells
QUESTION: The infection of plants by *Agrobacterium* involves all the following, EXCEPT:

A: it has been exploited for recombinant DNA approaches.
B: it involves a change in the plant genome.
C: it is prevented by secretion of acetosyringone.
D: it produces cells that can be grown indefinitely in culture.
E: it requires the presence of a wound.

Learning Response C: Correct. At the site of a plant wound where acetosyringone is being secreted, *Agrobacterium* transfers a plasmid, a part of which is inserted in the genome. This not only allows foreign genes to be introduced, but also produces indefinite growth in culture both of which are advantageous to its use in recombinant DNA techniques.

---

450. SUBJECT AREA: Special Features of Plant Cells
QUESTION: The following are true of plasmodesmata, EXCEPT:

A: they allow direct flow of molecules between cells.
B: they are roughly cylindrical.
C: they connect most cells.
D: they form cytoplasmic channels.
E: they only arise during cytokinesis.

Learning Response E: Correct. Plasmodesmata are membrane-lined cytoplasmic channels that connect most plant cells and allow direct flow of smaller molecules. They often form around smooth ER during cytokinesis, but can also arise de novo.

451. SUBJECT AREA: Special Features of Plant Cells
QUESTION: All the following are true of phloem EXCEPT:

A: it is in the symplast.
B: it primarily conducts sucrose to the leaves.
C: it represents an intercellular compartment.
D: its flow is driven by osmotic forces.
E: its principle conducting component is the sieve tube.

Learning Response B: Correct. The phloem represents a set of cells which are part of the intercellular compartment and conduct sugar by osmotic flow from its site of production (usually leaves) through sieve tubes to its site of use.

---

452. SUBJECT AREA: Special Features of Plant Cells
QUESTION: The following is true of xylem:

A: it is an intercellular compartment.
B: it requires evaporation.
C: its flow is bidirectional.
D: its structure is resistant to compression.
E: mineral salts and nitrogens diffuse into it.

Learning Response D: Correct. Xylem is composed of tubes with thick cell walls left by the death of cells. This prevents compression which allows flow of nutrients pumped in at the roots to proceed to the leaves driven by evaporation or root pressure.

---

453. SUBJECT AREA: Special Features of Plant Cells
QUESTION: Which of the following represents the most important difference between plant cells and the cells of other organisms?

A: absence of motility.
B: different defense mechanisms.
C: different intercellular communication.

D: different nutrition.
E: presence of a cell wall

Learning Response E: Correct. While all of these differences are important, it is the plant cell wall that is responsible for the majority of the other differences.

---

454. SUBJECT AREA: Special Features of Plant Cells
QUESTION: The secondary cell wall:

A: has a uniform composition.
B: is constructed to expand the primary cell wall and allow for growth.
C: is formed immediately after cell division.
D: is thin and semirigid.
E: provides more stability.

Learning Response E: Correct. The secondary cell wall is formed in cells, which have ceased to grow, by either thickening the existing primary wall or by deposition of a new wall of different material.

---

455. SUBJECT AREA: Special Features of Plant Cells
QUESTION: All the following characterize the primary cell wall, EXCEPT:

A: it has a constant composition.
B: it has a structure crosslinked by covalent and noncovalent interaction.
C: it is made up primarily of cellulose.
D: it is thin and semirigid.
E: its major components are well known.

Learning Response A: Correct. Primary cell wall is a thin and semirigid structure of variable composition. Its major components are known to consist of cellulose microfibrils crosslinked with other molecules by a variety of bond, both covalent and noncovalent.

---

456. SUBJECT AREA: Special Features of Plant Cells
QUESTION: Which of the following best describes hemicellulose?

A: it binds loosely to the surface of each cellulose microfibril.
B: it consists of linear polysaccharides.
C: it forms covalent bonds with cellulose.
D: it has a backbone of mixed carbohydrates.
E: it is divided into many classes.

Learning Response E: Correct. Hemicellulose is a heterogeneous group of branched polysaccharides that bind tightly but noncovalently to microfibrils and each other. While the specific sugars differ among species, they all have a backbone consisting of one type of sugar.

---

457. SUBJECT AREA: Special Features of Plant Cells
QUESTION: All of the following are true of turgor pressure in plants, EXCEPT:

A: it is directed outward.
B: it is driven by removing solutes from intercellular vacuoles.
C: it is maintained by the tensile strength of cell walls.
D: it is regulated by feedback mechanisms.
E: it provides the driving force for cell expansion during growth.

Learning Response B: Correct. Turgor pressure in plants provide the primary driving force for growth. A rupture in the cell wall allows a cell to expand irreversibly. It is a regulated process and following growth is retained by the pumping of solutes into the vacuoles causing their expansion due to osmolarity effects.

---

458. SUBJECT AREA: Special Features of Plant Cells
QUESTION: The formation of the secondary wall involves all the following, EXCEPT:

A: can involve deposition of lignin.
B: deposition opposite cytoplasmic microtubule bundles.
C: involves a change in the length of cellulose molecules.
D: is directed by nuclear events.
E: results in a more dense wall.

Learning Response D: Correct. The deposition of a secondary wall involves many changes. These appear to be cytoplasmically directed occurring opposite microtubule bundles. This involves construction of a more dense, less hydrated wall usually with longer cellulose molecules and often with the addition of associated molecules such as lignin.

---

459. SUBJECT AREA: Special Features of Plant Cells
QUESTION: Leaf abscission involves all the following, EXCEPT:

A: production of gas.
B: recycling of amino acids and ions to the plant.
C: removal of suberin.
D: response to a combination of ethylene and other regulators.
E: secretion of wall degrading enzymes.

Learning Response C: Correct. This process involves recovery of nutrients from the leaf. The production of ethylene gas in concert with other regulators then stimulates the release of pectinase and cellulase to digest cell walls. The stem side deposits suberin to protect the wound.

---

460. SUBJECT AREA: Special Features of Plant Cells
QUESTION: All of the following are characteristics of plastids, EXCEPT:

A: they are found in some plants.
B: they are unique to plants.
C: they develop from proplastids.
D: they have a double membrane.

E: they have their own genome.

Learning Response A: Correct. Plastids are a family of organelles that include chloroplasts, that are found in all plants. Developing from proplastids, they have a double membrane and their own genome.

---

461. SUBJECT AREA: Special Features of Plant Cells
QUESTION: Which of the following types of plastid is not involved in pigmentation?

A: amyloplasts.
B: chloroplasts.
C: chromoplasts.
D: etioplasts.
E: all of the following produce pigment.

Learning Response A: Correct. Amyloplasts, a form of leucoplasts, accumulate starch in cells. Etioplasts produce protochlorophyll, which, on their development into chloroplasts becomes chlorophyll, while chromoplasts produce carotenoid pigments.

---

462. SUBJECT AREA: Special Features of Plant Cells
QUESTION: Which of the following describes a tonoplast?

A: a cell wall.
B: a vacuole membrane.
C: a plastid.
D: a secretory vesicle.
E: a specialized plasmodesma.

Learning Response B: Correct. A tonoplast is the single membrane that separates a vacuole from the cytoplasm.

---

463. SUBJECT AREA: Special Features of Plant Cells

QUESTION: A plant vacuole can contain any of the following, EXCEPT:

A: opium.
B: rubber.
C: sodium.
D: trypsin inhibitors.
E: all of the above.

Learning Response E: Correct. Plants place various materials into their vacuoles. They use them for storage of proteins and molecules to be used in photosynthesis and growth, for osmotic regulation, and deposit substances and proteins to prevent attack by herbivores.

---

464. SUBJECT AREA: Special Features of Plant Cells
QUESTION: Which of the following is true of the production and deposition of cellulose?

A: it is deposited parallel to the direction of growth.
B: it is deposited perpendicular to cytoplasmic microtubules.
C: it is produced in the Golgi apparatus.
D: its synthesis has been studied *in vitro*.
E: its synthesis involves rosettes.

Learning Response E: Correct. Synthesis of cellulose occurs at the plasma membrane, at structures known as rosettes, and has not been reproduced *in vitro*. Cellulose is deposited parallel to the microtubules at the plasma membrane which run perpendicular to the direction of growth.

---

465. SUBJECT AREA: Special Features of Plant Cells
QUESTION: The following are true of the rotation of chloroplasts in response to light, EXCEPT:

A: it has been studied in *Mougeotia*.
B: it involves calcium-calmodulin interactions.
C: it involves rotation of the entire organelle.
D: it involves rotation perpendicular to dim light.

E: it involves several photoreceptors.

Learning Response C: Correct. In response to activation of phytochrome and blue-light receptors, the chloroplast will rotate perpendicular to dim light, and parallel to bright light. Studied in *Mougeotia*, the response has been found to involve calcium and calmodulin. Experiments with confined light sources have revealed that only that portion of the chlorophyll exposed to the light will rotate.

---

466. SUBJECT AREA: Special Features of Plant Cells
QUESTION: Which of the following describes the role of the suspensor?

A: it consists of maternal tissue.
B: it establishes the shoot apical meristem.
C: it gives rise to embryo.
D: it has dense cytoplasm.
E: it provides a pathway for nutrients.

Learning Response E: Correct. The suspensor forms from the first division of the zygote. It is vacuolated, and serves as a pathway between the maternal tissue and the embryo. It is the root apical meristem that faces the suspensor.

---

467. SUBJECT AREA: Special Features of Plant Cells
QUESTION: Which of the following classes of plant growth regulators is a gas?

A: abscisic acid.
B: auxins.
C: cytokinins.
D: ethylene.
E: gibberellins.

Learning Response D: Correct. While four of the classes of growth regulators are soluble hydrocarbons, ethylene functions as a gas.

---

468. SUBJECT AREA: Special Features of Plant Cells

QUESTION: The advantages of *Arabadopsis* for molecular studies include all of the following, EXCEPT:

A: a large number of clones have been isolated.
B: a cell culture system is possible.
C: it has a larger genome than most plants.
D: it has a short generation time.
E: it has a small size.

Learning Response C: Correct. Some of the advantages to *Arabadopsis* include its small size, short generation time, and small genome. It can be cultured, and individual mutations have easily been isolated.

---

469. SUBJECT AREA: Special Features of Plant Cells

QUESTION: The following are true of a callus, EXCEPT:

A: it can be induced to form a new plant.
B: it can be dispersed, producing totipotent cells.
C: it can proliferate indefinitely.
D: it represents a clonal population of cells.
E: its cells are relatively undifferentiated.

Learning Response D: Correct. A callus develops when part of a plant is placed in a nutrient culture. It appears undifferentiated, and can be propagated indefinitely or induced to form a new plant. While the cells can be dispersed, and are totipotent, they demonstrate a surprising degree of genetic variability.

---

470. SUBJECT AREA: Special Features of Plant Cells

QUESTION: The following are true of plant cell polarity, EXCEPT:

A: it has been studied in *Fucus*.
B: it is intrinsic to the zygote.
C: it is mirrored by an asymmetric distribution of ion channels.
D: it is regenerated by small lengths of stem.

E: it is stable.

Learning Response B: Correct. The polarity of a plant is stable and can be regenerated. As revealed by studies of *Fucus*, the zygote develops polarity of ion channels in response to some external gradient (gravity or light), and this appears to be responsible for the polarity of the developing plant.

---

471. SUBJECT AREA: Special Features of Plant Cells

QUESTION: Which of the following accurately describes the preprophase band?

A: it first appears just before interphase.
B: it is a subdivision of the cortical array.
C: it is formed from microtubules.
D: it is maintained throughout metaphase.
E: it is perpendicular to the newly forming cell wall.

Learning Response C: Correct. After the disappearance of the cortical array at the end of interphase, the preprophase band appears, but disappears again prior to metaphase. Even though it is no longer present when the cell wall is formed, it marks the exact location where it will form.

---

472. SUBJECT AREA: Cancer

QUESTION: Tumors are usually classified based on all of the following, EXCEPT:

A: cell type giving rise to the tumor.
B: its ability to invade other tissue.
C: its presence or absence in secondary tissues.
D: the nature of the genetic mutation.
E: tissue type giving rise to the tumor.

Learning Response D: Correct. Tumors can be classified based on any or all of these factors. However, the research necessary to determine the nature of the genetic abnormality is usually not performed, and so the nature of the mutation giving rise to a tumor is generally not known.

473. SUBJECT AREA: Cancer
QUESTION: Location of a metastasis indicates:

A: that the tumor can invade surrounding tissue
B: that the tumor can undergo normal metabolism
C: that the tumor has spread to other locations
D: that the tumor is benign in nature
E: the tumor's tissue of origin

Learning Response C: Correct. A metastasis is a secondary tumor which has arisen when cells from a primary malignancy have traveled through the bloodstream to a new location in the body, forming a new tumor there.

474. SUBJECT AREA: Cancer
QUESTION: The following would be expected of a testicular tumor EXCEPT:

A: any DNA abnormality would be shared by all of the cells.
B: the cells are all of the same testicular cell type.
C: the cells would all have the same X chromosome inactivated.
D: the tumor would have a single focus, prior to metastasis.
E: all of the above would be expected.

Learning Response C: Correct. While these are general characteristics of tumors, X-inactivation only occurs in females, and thus would not be found in a testicular tumor.

475. SUBJECT AREA: Cancer
QUESTION: The Ames test, performed in *Salmonella* to test for mutagenicity is an effective mechanism for predicting the carcinogenic potential of a compound because:

A: carcinogenesis is due to mutagenesis, the mechanisms for which are essentially the same in any organism.

B: it is not known why this works, but the correlation has proven true in all studies.
C: mutation of the liver extract stimulates growth of the Salmonella.
D: Salmonella are often responsible for tumors, especially of the liver.
E: Salmonella develop tumors similar to those seen in humans when incubated in the presence of an activated liver extract

---

Learning Response A: Correct. This assay measures the ability of the mutagen to cause mutations in the *Salmonella*, which should be proportional to its ability to cause mutation, and hence its carcinogenic potential in animals. The liver extract modifies the mutagen, potentially activating it, as normally occurs in the liver.

---

476. SUBJECT AREA: Cancer
QUESTION: The fact that a human's risk for cancer increases progressively with age indicates that:

A: cancers grow slowly, and it required decades for them to reach a detectable size.
B: humans develop tumors more easily than mice.
C: multiple mutations are required to produce cancer.
D: mutations in the genome do not cause cancer.
E: mutations occur rarely.

Learning Response C: Correct. This result demonstrates that for a tumor to develop, multiple mutations are necessary. Were this not the case, then one would expect the rate to be constant (or to level out at a constant risk after the initial incubation time had expired for a slow growing tumor). Instead, the progression of risk is logarithmic, indicating that it is the sum of the probabilities of several mutations accumulating in the same cell.

---

477. SUBJECT AREA: Cancer
QUESTION: Analysis of the progression to leukemia and cervical cancer tells us that tumors:

A: are easier to cure in their final stages.
B: become more differentiated as they progress.
C: convert the cells around them to tumorous tissue.
D: pass through successively more abnormal stages.
E: proliferate unchanged from the time of their origin.

Learning Response D: Correct. Tumor growth naturally selects for the more rapidly proliferating cells. Thus, successive aberrations in the cells which increase their growth potential cause the tumors to become progressively more proliferative, less differentiated, and more difficult to cure.

---

478. SUBJECT AREA: Cancer
QUESTION: The selection for more severe cancer phenotypes will involve all of the following, EXCEPT:

A: the degree of contact inhibition remaining.
B: the rate at which mutations take place.
C: the rate of cell division.
D: the total size of the tumor.
E: all of the above would play a role.

Learning Response E: Correct. The selection of more severe phenotypes involves all of the characteristics of evolution of populations, including reproductive potential (rate of division), mutation rate (particularly for gross rearrangements), the selective advantage (as revealed by the degree of contact inhibition), and the total tumor size, effecting the total potential for mutations to occur (since rate is mutations per cell).

---

479. SUBJECT AREA: Cancer
QUESTION: Which of the following statements about tumor initiators and promoters is true?

A: a tumor initiator cannot cause a tumor, but requires a promoter.
B: a tumor promoter will only affect an initiator-exposed cell within a narrow window of time.
C: all cancer causing agents are mutagenic.

D: tumor promoter activity often depends on a threshold exposure.
E: wounding can be a tumor initiator.

Learning Response D: Correct. A tumor initiator such as a mutagen, while capable of causing a tumor by themselves, can also create mutations in the DNA which might be unmasked by a threshold exposure to a promoter at any time in the future. A promoter may be any substance or factor (such as wounding) which alters the expression of the cell, unmasking the effect of the initiator.

---

480. SUBJECT AREA: Cancer
QUESTION: Which of the following accurately portrays cancer epidemiology?

A: immigrants resemble the group from which they came in their susceptibility to a particular cancer.
B: specific cancers appear at relatively equal frequencies independent of the location.
C: studies begin by discovering the mechanism by which a factor increases cancer risk.
D: the risk of a specific cancer is strongly influenced by local factors.
E: there are vastly different total cancer risks for different geographic areas.

Learning Response D: Correct. The risk for a particular cancer is highly dependent on the individual factors one is exposed to, which differ tremendously from place to place. This is proven by the similar patterns seen in natives and immigrants exposed to the same environment. However, the total frequency appears to differ little, indicating that each area has different risks for different cancers, and these average out.

---

481. SUBJECT AREA: Cancer
QUESTION: The difficulties involved in developing a cure for cancer include all of the following, EXCEPT:

A: it is difficult to remove every cancer cell through surgery.
B: selectivity of a treatment toward the tumor is difficult to attain.
C: there are no cures to be used as guides for development of future drugs.
D: toxicity to normal cells precludes aggressive treatment of tumors.
E: tumor cells can develop drug resistance.

Learning Response C: Correct. While there are many problems with identifying cancer treatment protocols, fortunately there have been some successes, and while it is still a trial and error process, the experience with the cures that have been developed at least provide the methodologies by which other tumors can be addressed.

---

482. SUBJECT AREA: Cancer
QUESTION: Which of the following characteristics of metastatic cells is likely to be the most difficult for them?

A: their ability to break through the basal lamina.
B: their ability to enter the circulation.
C: their ability to loosen their attachment to other tumor cells.
D: their ability to migrate through other tissue.
E: their ability to thrive in their new environment.

Learning Response E: Correct. Probably most difficult for metastatic cells is the ability to thrive in a new tissue environment. They have specifically been selected for growth in the tissue of origin. Thus they may not find the growth conditions in a new tissue permissive.

---

483. SUBJECT AREA: Cancer
QUESTION: Why are defects in replication and repair often found in tumors?

A: these mutations in replication cause the cells to proliferate.

B: they make the cells more liable to suffer additional mutations.
C: they prevent cells from differentiating.
D: they represent the cell's attempt to mutate the tumor into inactivity.
E: tumor development is known to cause such mutations.

Learning Response B: Correct. Mutations in the DNA repair, replication, and recombination increase the frequency at which other genes are mutated, and thus increase the probability of a tumor developing.

---

484. SUBJECT AREA: Cancer
QUESTION: Multidrug resistance is the result of:

A: a viral oncogene.
B: amplification of a specific chromosomal region.
C: many individual drug resistance genes.
D: the dihydrofolate reductase gene.
E: treatment regimens using more than one drug.

Learning Response B: Correct. The development of multidrug resistance results from the amplification of a gene for a membrane transport ATPase which is thought to export a broad range of drugs, and can be induced to be amplified by any of them.

---

485. SUBJECT AREA: Cancer
QUESTION: A proto-oncogene is:

A: a gene carried by a retrovirus and which causes cancer.
B: a gene which serves as a tumor suppressor, when both copies are mutated.
C: a gene which is only expressed by a tumor.
D: a mutated cellular gene which causes tumors.
E: the normal allele of a gene which causes tumors when one copy is mutated.

Learning Response E: Correct. All oncogenes discovered so far have a normal cellular counterpart — a proto-oncogene — which has no oncogenic activity. Proto-oncogenes can mutate to form oncogenes.

---

486. SUBJECT AREA: Cancer

QUESTION: Which of the following statements about cancer-causing viruses is true?

A: DNA and RNA viruses are responsible for all tumors.
B: DNA viruses carry with them copies of cellular genes.
C: RNA viruses require special enzymes to replicate.
D: only RNA viruses can cause cancer in humans.
E: they can only cause a tumor if they carry an oncogene.

Learning Response C: Correct. While both DNA and RNA viruses can cause cancer, they do this by different mechanisms. DNA viruses carry foreign genes which subvert the normal cellular control mechanisms. RNA viruses often carry mutated copies of normal cellular genes, as well as their own reverse transcriptase, which is necessary for replication. Finally either can cause tumors through insertion in or near a proto-oncogene.

---

487. SUBJECT AREA: Cancer

QUESTION: In tumors oncogenes have been identified which derive from all of the following steps, EXCEPT:

A: a hormone.
B: a nuclear receptor.
C: a signal transduction protein.
D: a tyrosine kinase receptor.
E: all of the above are examples of proto-oncogenes.

Learning Response E: Correct. Nearly every step in the cell signaling/regulation system has provided examples of proto-oncogenes. This includes hormones (*sis*), tyrosine kinase receptors (*erb-B*), signal transduction (*ras*), and DNA binding proteins (*erb-A*, a nuclear receptor; and *fos*).

488. SUBJECT AREA: Cancer

QUESTION: Insertional mutagenesis works through:

A: the insertion of a transposon into a virus.
B: the insertion of viral enhancers near a proto-oncogene.
C: the insertion of an oncogene into the gag gene.
D: the insertion of an oncogene into the genome.
E: the insertion of an oncogene into the virus.

Learning Response B: Correct. Insertional mutagenesis is the result of the strong viral enhancers being juxtaposed with the promoter of a proto-oncogene. This brings about abnormal levels of expression, which alters the normal control of the cell.

---

489. SUBJECT AREA: Cancer

QUESTION: Approaches which have resulted in the discovery of oncogenes include all of the following EXCEPT:

A: the analysis of sequences near the site of viral insertion.
B: the investigation of chromosomal translocations.
C: the sequencing of tumor causing viruses.
D: the transfection of cells with DNA from tumors.
E: all of the above approaches have been used.

Learning Response E: Correct. All of the methods indicated have been used to identify oncogenes.

---

490. SUBJECT AREA: Cancer

QUESTION: Alterations which cause proto-oncogenes to become oncogenes:

A: can include point mutations outside of the coding region.
B: must mutate the coding region of the protein.
C: require the presence of a virus.

D: require mutation to occur in the process of gene amplification.
E: result in an altered gene product.

Learning Response A: Correct. The process of converting a proto-oncogene to an oncogene can involve any type of mutation or rearrangement which changes the level of activity of the protein. This includes mutations in the protein which change its activity, and mutations or amplification, which change the total amount of protein being expressed in a cell.

---

491. SUBJECT AREA: Cancer
QUESTION: When injecting an oncogene into a mouse to create a transgenic, one would expect:

A: no viable progeny that carry the gene.
B: the mother to develop tumors from the embryos.
C: tumors in every cell of the progeny which carry the gene.
D: tumors in every mouse which carries the transgene.
E: tumors only in cells which accumulate other mutations.

Learning Response E: Correct. Since the development of tumors is thought to require multiple mutations, one would expect to see tumors develop only in mouse cells which have independently acquired mutations of other proto-oncogenes. In addition, because of the nature of transgene expression, there may be mice which fail to express the transgene, and hence never develop tumors.

---

492. SUBJECT AREA: Cancer
QUESTION: The fusion of a transformed cell with a normal one indicates the presence of tumor suppressors because:

A: mutations in only one copy of a tumor suppressor are necessary for cancer.
B: the fusion behaves like the transformed cell, unless there is a mutation in the suppressor.
C: the fusions never become cancerous.

D: the loss of the transformed phenotype indicates that a suppressor has come from the normal cell.
E: this process results in mutations in tumor suppressor genes, which can then be studied.

Learning Response D: Correct. In these experiments, the fusions have a normal phenotype, unless they lose a chromosome containing a tumor suppressor from the fusion. This chromosome would have come from the normal cell, and has this effect because only one normal copy of the suppressor gene is required for suppressor activity.

---

493. SUBJECT AREA: Cancer
QUESTION: In retinoblastoma,

A: a process occurs which is not seen in other tumors.
B: those inheriting a mutated gene develop tumors in every retinal cell.
C: tumors contain alterations in both copies of the *Rb1* gene.
D: tumors only occur in people who inherit one defective gene.
E: tumors only result from chromosomal rearrangements.

Learning Response C: Correct. In retinoblastoma, as with a few other rare tumors, a mutation in both copies of a tumor suppressor gene results in the cells no longer being appropriately regulated. Thus a person who has inherited one faulty copy requires only a mutation or other alteration in the other copy to develop a tumor, while a normal individual requires mutations to occur in both.

---

494. SUBJECT AREA: Cancer
QUESTION: Studies of lung cancer tumors:

A: demonstrate the same oncogene in all tumors.
B: have not been extensive enough to draw conclusions.

C: never show chromosomal abnormalities seen in other tumors.
D: reveal an alteration in the same gene in almost all tumors.
E: none of the above are true.

Learning Response D: Correct. These tumors have been extensively studied, and reveal a wide range of oncogenes being involved, as well as observable chromosome abnormalities in these tumors. However, when a probe directed against a region of chromosome 3 is used to study the DNA of the tumor, alterations are seen in almost every case, indicating that an important suppressor gene resides in this area.

---

495. SUBJECT AREA: Cancer
QUESTION: The following chromosomal abnormalities can be revealed when karyotyping tumors, EXCEPT:

A: deletion of regions of the chromosome.
B: double minute chromosomes.
C: chromosome translocations.
D: homogeneous staining regions.
E: viral insertions.

Learning Response E: Correct. While all of the above can be responsible for transformation, they cannot all be seen when karyotyping. Some deletions and all viral insertions are too small to be seen when karyotyping, and these require a technique with higher resolution, such as RFLP analysis.

Index

Numbers after index entries represent question numbers.